Mathematical Engineering

Today, the development of high-tech systems is unthinkable without mathematical modeling and analysis of system behavior. As such, many fields in the modern engineering sciences (e.g. control engineering, communications engineering, mechanical engineering, and robotics) call for sophisticated mathematical methods in order to solve the tasks at hand.

The series Mathematical Engineering presents new or heretofore little-known methods to support engineers in finding suitable answers to their questions, presenting those methods in such manner as to make them ideally comprehensible and applicable in practice.

Therefore, the primary focus is—without neglecting mathematical accuracy—on comprehensibility and real-world applicability.

To submit a proposal or request further information, please use the PDF Proposal Form or contact directly: Dr. Thomas Ditzinger (thomas.ditzinger@springer.com)

Indexed by SCOPUS, zbMATH, SCImago.

Mark W. Mueller

Dynamics and Control of Autonomous Flight

Springer

Mark W. Mueller
Department of Mechanical Engineering
University of California
Berkeley, CA, USA

ISSN 2192-4732 ISSN 2192-4740 (electronic)
Mathematical Engineering
ISBN 978-3-031-91870-4 ISBN 978-3-031-91871-1 (eBook)
https://doi.org/10.1007/978-3-031-91871-1

This Springer imprint is published by the registered company Springer Nature Switzerland AG
The registered company address is: Gewerbestrasse 11, 6330 Cham, Switzerland

Preface

This book is built from my lecture notes for the UC Berkeley Mechanical Engineering course ME136/ME236U, "Dynamics and Control of Autonomous Flight". I originally conceived the course as the answer to the question "what do I need to teach someone so that they can work productively in my research group?".

I assume that the reader is familiar with engineering fundamentals, linear algebra, differential equations, and the like. We assume no particular knowledge of fluid mechanics, of dynamics, or of controls—however, because the topic is so broad, having prior knowledge of topics will allow the student to better place in context the particular aspects of (e.g.) controls that we use in this course.

A large part of this material is developing the mathematical tools to *clearly* model and describe three-dimensional mechanics. For this, we follow the excellent tensor-based approach of Zipfel (2007). A characteristic of this approach is that the notation often feels somewhat verbose and "heavy" to students, who would often like to take a shortcut and "get to the flying stuff quicker". I make a point of doing this slowly, and thoroughly, as the notation encodes a lot of information about the underlying phenomena, and being able to use it well will give the engineer greater insight into the system, and more easily spot mistakes in their own derivations (and, from experience, it is very easy to make a mistake when dealing with 3D mechanics).

Throughout this book, we will use the quadcopter as our motivating example. The intention is to anchor ideas in a particular system, for which the student can then develop deep intuition; the resulting connections will then ideally lead to a deeper understanding of the various concepts in this book.

When I use this book to teach the course at Berkeley, the theory course is accompanied by two optional laboratory courses: the first uses low-cost, lightweight quadcopters, and the second is done fully in simulation. There are then six big labs that students work through in the semester: (1) setting up the required software, and becoming familiar with the C++ coding environment; (2) identifying the characteristics of the powertrain; (3) attitude estimation; (4) attitude control; (5) hover control;

and (6) a final competition. Anyone interested in offering a similar version of this course is encouraged to contact me.

Berkeley, CA, USA Mark W. Mueller

Reference

Zipfel PH (2007) Modeling and simulation of aerospace vehicle dynamics, Second Edition. AIAA

Competing Interests The author has no competing interests to declare that are relevant to the content of this manuscript.

Contents

1 Fundamentals of Flight 1
1.1 Fundamental Forces of Flight 1
1.1.1 Fluid Mechanics, in a Nutshell 2
1.1.2 Aerodynamic Forces: Lift and Drag 5
1.1.3 Aerodynamic Forces: Propeller Thrust 9
1.2 Energy and Propulsion 10
1.2.1 Powertrain 11
1.2.2 Power Consumption in Flight 12
1.3 Actuators and Control Surfaces 14
Problems for Chapter 1 15
References 19

2 Three-Dimensional Geometry 21
2.1 Nomenclature 21
2.1.1 Invariant Description 21
2.1.2 Coordination and Coordinate Systems 22
2.2 Tensors 27
2.2.1 Properties of Tensors 27
2.2.2 Algebra of Tensors 28
2.3 Frames and Coordinate Systems 31
2.3.1 Coordinate Transformation Matrix 32
2.3.2 Euler angles 34
2.4 Rotation Tensor 35
Problems for Chapter 2 37
References 47

3 Kinematics 49
3.1 Rotational Time Derivative 49
3.2 Translational Velocity and Acceleration 51
3.3 Angular Velocity 52
3.3.1 Orientation Differential Equations 53

3.3.2 Euler Transformation 55
Problems for Chapter 3 56
Reference 58

4 Dynamics of Rigid Bodies 59
4.1 Dynamics of Translation 59
4.1.1 Translational Momentum 60
4.1.2 Newton's Law of Motion 60
4.2 Attitude Dynamics 61
4.2.1 Inertia Tensor 61
4.2.2 Angular Momentum 62
4.2.3 Moments 62
4.2.4 Euler's Law 63
4.3 Coordination of Attitude Dynamics 63
4.4 Lagrangian Mechanics 64
Problems for Chapter 4 64
Reference 69

5 Sensing and Estimation 71
5.1 Sensor Basics 71
5.2 Some Key Sensors 72
5.2.1 Accelerometers 73
5.2.2 Rate Gyroscopes 75
5.2.3 Global Navigation Satellite Systems (GNSS) 77
5.2.4 Barometer 79
5.2.5 Magnetometer 79
5.2.6 Range Sensors 80
5.2.7 Optical Flow 81
5.2.8 Cameras, Radar, and Lidar 82
5.3 Estimation 82
5.3.1 Tilt Estimation 83
5.3.2 Estimating the Tilt and Height 86
5.3.3 Estimating the Tilt, Height, and Velocity 87
5.3.4 Estimating Position with GPS 88
5.3.5 Visual-Inertial Odometry (VIO) 88
Problems for Chapter 5 88
References 92

6 Quadcopter Dynamics 93
6.1 Propeller Forces and Torques 94
6.2 Transformation of the Inputs 97
6.3 State-Space Description 98
6.3.1 Equilibrium and Linearization About Hover 99
Problems for Chapter 6 101

7 Control and Stabilization 105
7.1 Language, Symbols, and Equations 105
7.1.1 Equilibria 106
7.1.2 Linear Versus Nonlinear Systems 107
7.2 Control of Simple Systems 107
7.2.1 Single Integrator 108
7.2.2 Double Integrator 109
7.2.3 Feedforward Signals 111
7.2.4 Cascaded Control and More Complex Systems 112
7.2.5 Avoiding Cascaded Control 114
7.3 Stability 115
7.3.1 Stability in the Sense of Lyapunov 115
7.3.2 Testing for Stability: Lyapunov's Indirect Method 118
7.4 Other General Control Approaches 120
7.5 A Quadcopter Control Architecture 121
7.5.1 Angular Velocity Control 122
7.5.2 Attitude Control 122
7.5.3 Position Control 123
7.5.4 Acceleration Transform 123
7.6 Trajectory Generation and Tracking 124
Problems for Chapter 7 127
References 134

8 Advanced Topics 135
8.1 Other Attitude Parametrizations 135
8.1.1 Rotation Vectors and Axis-Angle Representation 136
8.1.2 Euler Symmetric Parameters (ESP) 139
8.1.3 Comparison of Attitude Representations 142
8.2 Actuator Disk Model of Propeller Thrust and Power 143
8.2.1 Bernoulli's Law 143
8.2.2 Derivation 144
8.3 Principled Model-Based State Estimation 146
8.4 The Dynamics of Unforced Rotations 148
Problems for Chapter 8 153
References 159

Appendix: The Standard Atmosphere 161

Index 163

Chapter 1
Fundamentals of Flight

You're the wind beneath my wings.

—Bette Midler

Our objective in this chapter is to describe the fundamental physics of heavier-than-air flight. We aim to understand, at a high level, how airflow can be used to create forces, and how we harness those forces for flight. This will be done at a high level only, as the course lacks enough time to go deeper (and, in fact, you'd need many courses if you wanted to really go 'deep'!). A good reference for aerodynamics is the book by McCormick (1995), or that of Anderson (2011). This chapter will describe both fixed-wing flight, and propeller-driven (rotary-wing) flight; later though we will be primarily concerned with propeller-driven aircraft.

1.1 Fundamental Forces of Flight

The most important force for any heavier-than-air craft is its weight. For a typical fixed-wing aircraft moving horizontally, the force that opposes the weight is called the *lift*. To produce lift, the aircraft interacts with the surrounding air, and specifically imparts momentum to the air to experience a reaction force. In this process of imparting momentum, the aircraft also experiences a force opposing its motion, which is called the *drag*. Thus, to maintain its constant speed, a *thrust* force is required to oppose the drag force. To a first, crude, approximation, a fixed-wing aircraft in steady flight will have the lift equal the weight, and the thrust equal the drag, as is shown in Fig. 1.1. Of course, in practice, the thrust vector may not be perfectly horizontal in flight, so that a component of thrust is used to balance the vehicle weight.

M. W. Mueller, *Dynamics and Control of Autonomous Flight*, Mathematical Engineering,
https://doi.org/10.1007/978-3-031-91871-1_1

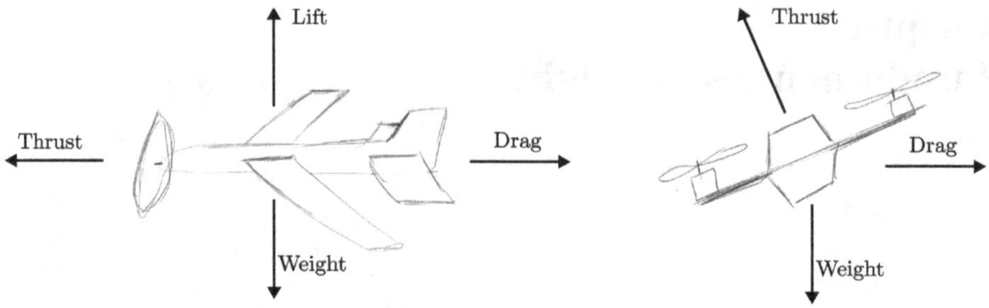

Fig. 1.1 A simplified force balance for steady flight of a fixed-wing aircraft (left) and a multicopter (right), in steady flight

For rotorcraft, such as a multicopter, there is typically no (or negligible) lift, so that the force balance in steady flight is much simpler. Instead, the vehicle's thrust is tilted forward, so that a component of the thrust is used to balance the vehicle drag, and a (typically larger) component balances the vehicle weight.

In this chapter, we hope to build an understanding for how these forces are generated, and how they relate to the system characteristics. We will then eventually use these models to also make predictions about non-steady flight, for example when an aircraft is maneuvering. This will also be crucial for understanding the design tradeoffs of an aircraft, including how energy is converted into forces, so that we may design efficient aircraft.

1.1.1 Fluid Mechanics, in a Nutshell

We start with a review of fluid mechanics concepts, restricting ourselves to the study of air. To model a fluid, we define two parameters: the *density* of the fluid ρ, which captures the amount of mass per unit volume (usually in [kgm^{-3}]). Density is important, because aerodynamics requires pushing air around, and the denser the air is the more force is required to 'push the air around'. The second parameter is the *viscosity*, μ, of the fluid, a measure of its resistance to shear deformation, and commonly with units [Pa s].

A definition of viscosity comes from considering two large, flat plates, separated by a small fluid-filled gap of width y (see Fig. 1.2). The plates are moved relative to one another at a speed of v, and the viscosity μ relates the sheer pressure (that is, the force f required to keep the plates moving at a constant speed, normalized by the area A of the plates) to the velocity gradient:

$$\frac{f}{A} = \mu \frac{v}{y} \tag{1.1}$$

Our intuition of a fluid's 'thickness' (as in, 'honey is thicker than water') is related to the fluid's viscosity.

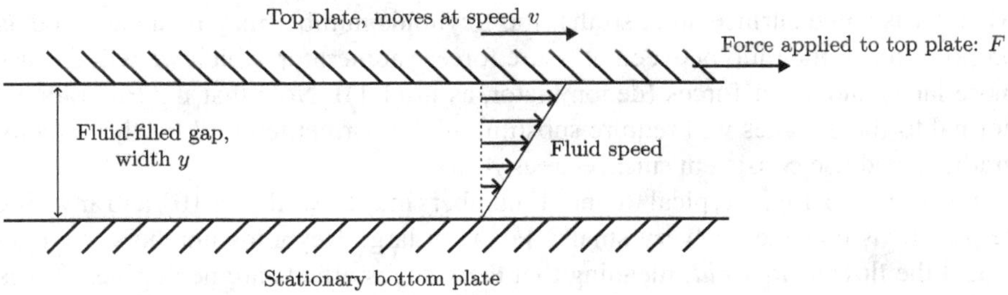

Fig. 1.2 The shear force F applied to move two parallel plates with a narrow gap y at constant relative speed v is related to the viscosity μ of a fluid, per (1.1). Note that the fluid's speed varies linearly through the gap

In 'the standard atmosphere' (idealized conditions at sea level), we use a value of $\rho = 1.225\,\mathrm{kgm}^{-3}$ for air. Water has a density three orders of magnitude higher, at approximately $1000\,\mathrm{kgm}^{-3}$. For the viscosity of air we use $\mu = 18.7 \times 10^{-6}\,\mathrm{Pa\,s}$, see Appendix A for values under differing conditions.

The characteristics of a fluid's flow under any particular condition can be exceedingly complex. However, several 'dimensionless numbers' have been developed that aim to characterize certain aspects of a flow, typically so that we may be confident in extrapolating from a small-scale experimental setup to a larger-scale system, as long as we can ensure that the corresponding characteristics (i.e., dimensionless numbers) match. The two most important such numbers are the *Reynolds number* and *Mach number*, each named after scientists who popularized, or invented, them.

Reynolds Number

The Reynolds number is typically written as *Re*, and is defined for a steady flow as

$$Re = \frac{v_\infty l \rho}{\mu} \tag{1.2}$$

wherein v_∞ is the *free-stream velocity* (that is, the relative speed of the airflow far ahead of the object interacting with it), l is some *characteristic length*, and ρ and μ the fluid density and viscosity. The characteristic length, l, is defined by the application, and one must therefore be careful when comparing Reynolds numbers from different sources (are they consistent with the length they use?). For example, when considering the airflow over a sphere (or golf ball), we use the diameter as the characteristic length, when looking at an airfoil the cord, and when considering a full aircraft typically the wingspan. Thus a Reynolds number is meaningless if it is not also clearly provided what the characteristic length is that was used.

With a bit of effort, Eq. (1.2) can be manipulated to

$$Re = 2\frac{\frac{1}{2}\rho v_\infty^2 A}{\mu \frac{v_\infty}{l} A} \tag{1.3}$$

where A is some surface area, so that the Reynolds number may be understood as proportional to the ratio between pressure forces (numerator, which we will discuss more later) and sheer forces (denominator, as in (1.1)). Note that the true ratio of normal to shear forces will require substituting the correct length l for the velocity gradient, and use consistent surface areas A.

For aircraft in flight, typical Reynolds numbers may be well over 10^6, and anything below 10^5 is considered 'very small'. For very large Reynolds numbers, we may model the flow as *inviscid*, meaning that the viscous effects are negligible. This is desirable, as it allows us to make significant simplifications—we will thus often assume inviscid flow.

Note—The Reynolds number is typically a large number, and it is tempting to communicate it in all its glory. However, this is not useful; the Reynolds number is indicative as an order-of-magnitude value, and anything beyond two significant digits should be considered 'noise', and not reported (i.e., a good way of reporting it may be something like $Re \approx 1.5 \times 10^6$, and certainly not $Re = 1480347.02\ldots$).

Mach Number

Our second dimensionless number is the Mach number, M, defined as

$$M := \frac{v_\infty}{v_{\text{sound}}} \tag{1.4}$$

or the ratio between the free stream velocity v_∞ and the speed of sound in the fluid v_{sound}.speed of sound For the air at sea level, this is approximately $v_{\text{sound}} \approx 343\,\text{m}\,\text{s}^{-1}$ (see Appendix A). As the flow speed approaches the speed of sound, complex compressibility effects radically change the flow behavior. We refer to flow at small Mach numbers as *incompressible*, and can assume that the flow density is constant throughout, allowing for significant simplifications in analysis.

Two asides here: first, our intuition about aerodynamics often comes from looking at birds and fish (which operate at extremely low Mach numbers). For this reason, we typically perceive machines designed for low Mach numbers as "sleek", while machines designed for supersonic (or hypersonic) flight appear "ugly", even though they are all engineered ruthlessly for efficiency.[1] Secondly, while the air is certainly more compressible than water, assuming the air is incompressible is not that outlandish (have you ever tried to compress a balloon in your hands?).

[1] Presumably, our lizard brains see something that looks like a swooping eagle and thinks 'that must be fast and agile', but sees something that looks like the space shuttle (and is designed for hypersonic flight) and sees something that looks relatively blunt and crude.

1.1.2 Aerodynamic Forces: Lift and Drag

We now turn our attention to characterizing the steady flow of a fluid over a surface, and how this creates aerodynamic forces. We start by considering a two-dimensional object, an airfoil, before we consider a three-dimensional wing, and an entire aircraft.

Aerodynamic forces are typically modelled using dimensionless coefficients, in the form below for some force f_L:

$$f_L = \frac{1}{2}\rho v_\infty^2 A C_L \tag{1.5}$$

where ρ is the fluid density, v_∞ is the relative speed of the free-stream flow (i.e., far ahead of the surface), A is some representative surface area, and C_L is a dimensionless coefficient. The term $\frac{1}{2}\rho v_\infty^2$ is called the 'dynamic pressure', often replaced with the symbol q—for example, live mission-control commentary on rocket launches will often call out "Max q" when the rocket passes through the region of greatest aerodynamic loading (that is, maximum dynamic pressure).

The coefficient C_L must specify the flow regime for which it is valid (Reynolds and Mach numbers), as well as what the reference area A is to be used. The subscripts of force and coefficient (here, L) denote the forces used. By convention, we use lower case subscripts (e.g., C_l and C_d) for two-dimensional flows, and upper case symbols (like C_L and C_D) for three-dimensional flows.

Flow over a 2D Airfoil

An airfoil is a cross-section of a wing, and studying the flow over an airfoil is equivalent to the flow over an infinitely long wing (where there is no flow along the wing's length). The two-dimensional case is significantly simpler than the flow over a 3D surface, but still allows us to make useful predictions. We are typically primarily interested in the lift and drag produced by an airfoil in a flow, typically as the lift (C_l) and drag (C_d) coefficients, which are related to the forces through (1.5). These can be determined experimentally (typically an expensive process using a finely calibrated wind tunnel) or through numerical simulation (either an expensive process using high-fidelity computational fluid mechanics, or a less accurate and less expensive approximation tool).

Results are typically gathered as a function of the *angle of attack*, the angle between the free stream velocity and a line connecting the leading and trailing edges of the airfoil. The angle of attack is typically denoted as α, and illustrated in Fig. 1.3.

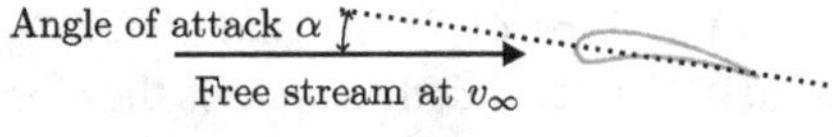

Fig. 1.3 Definition of the angle of attack of an airfoil

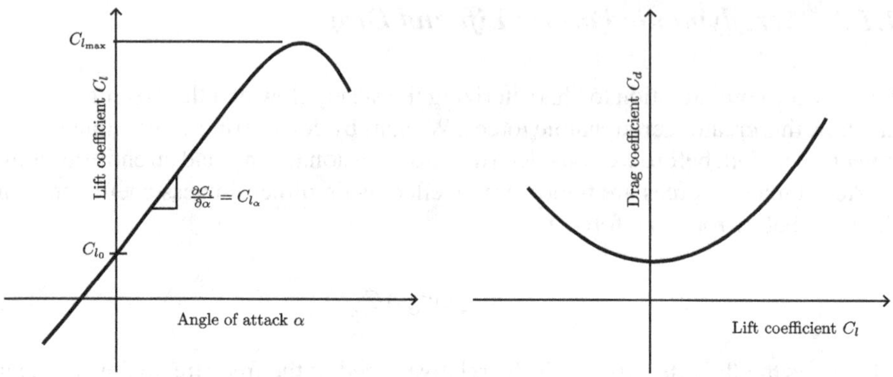

Fig. 1.4 Typical data of an aerofoil lift/drag characteristics

A typical result is a set of curves of lift and drag coefficients, as in Fig. 1.4. The figure on the left shows how the lift coefficient changes with angle of attack, and shows three important characteristics:

- the lift coefficient at zero angle of attack, C_{l_0};
- the slope of the lift coefficient for small angles of attack, $C_{l_\alpha} = \frac{\partial C_l}{\partial \alpha}$; and
- the maximum lift coefficient, $C_{l_{max}}$.

For small angles of attack, we can thus approximate the lift coefficient as below.

$$C_l \approx C_{l_0} + C_{l_\alpha}\alpha \tag{1.6}$$

As the angle of attack keeps increasing, this affine relationship breaks down, until we reach the maximum lift coefficient $C_{l_{max}}$, beyond which the lift decreases. This is where the airfoil is *stalled*, and the flow no longer follows the shape of the airfoil well.

The second plot in Fig. 1.3 shows the drag coefficient as a function of the lift coefficient. We see that drag is positive, even for negative lift (of course, the flow always opposes the wing). We expect a dramatic increase in drag after stall.

Where the flow may be assumed to be inviscid (i.e., as $Re \to \infty$ and effects from viscosity become relatively insignificant) the 2D flow over a thin airfoil can be approximated using very simple analytic results (see 'Thin Airfoil Theory', e.g., McCormick (1995, Chap. 3)) which replaces the airfoil surface with 'vortices', and calculates the resulting change in fluid momentum (i.e., lift force).

We will not derive the results here, instead only state the main results. For a 'flat plate' (i.e., an airfoil that is a thin, straight line) the lift generated can be modelled as with the slope $C_{l_\alpha} = 2\pi$ and offset $C_{l_0} = 0$ (with the slope requiring the angle be in [rad]). For curved, thin airfoils the slope remains the same, but the offset will be non-zero. A surprising result for these inviscid flows over thin airfoils is that zero drag is predicted—in reality, the lift prediction is often a reasonably first approximation, though the drag is obviously not. We will see, though, that for finite (3D) wings, a component of drag can be predicted with similar ease.

Flow Over a 3D Wing

Three dimensional flow is significantly more complex than that over a simple airfoil. Making accurate predictions about aircraft performance typically requires high-fidelity computational fluid dynamics simulations, validated against wind-tunnel test results. However, some easy predictions of major effects can be made more simply.

A major feature of a finite wing is the wing-tip. When producing lift, the pressure below the wing is higher than that above the wing (integrating this pressure difference over the wing surface is, of course, the lift force). For infinite (2D) wings, the air has no choice but to push up on the wing, producing lift. For 3D wings, some fraction of the air escapes around the wing-tip from the high-pressure region below the wing to the low-pressure region above the wing. This modification in airflow reduces the lift that a real wing produces, compared to that of an infinite (2D) wing; simultaneously, it increases the drag the wing experiences. This is shown in the photos of Fig. 1.5.

We first characterize a wing by its *wingspan* (b), surface area (S), and aspect ratio (typically written as $A\!\!\!/R$), illustrated in Fig. 1.6. The wingspan is the distance, seen from the top, between the two wing-tips, while the surface area is the projected area from above the wing (that is, the area of the shadow it casts, not the total surface area around the wing). We compute the aspect ratio as

Fig. 1.5 Flow over a finite wing tends to curl up around the wing tips, leading to a reduction in lift and increase in drag. These two images are from wingtip studies by NASA

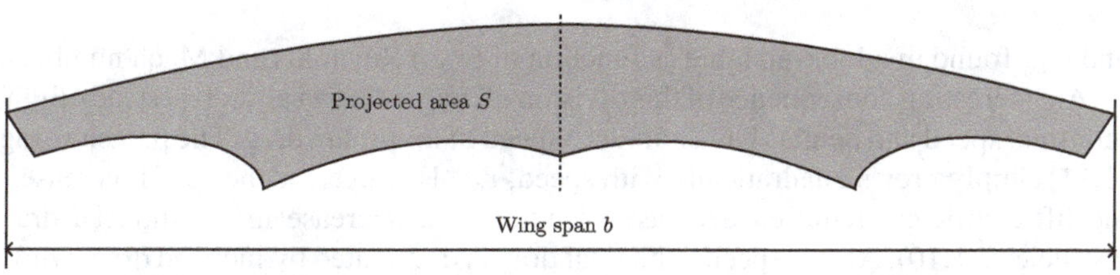

Fig. 1.6 The view from directly above of a (strange) wing, showing the projected surface area and wing span, from which the aspect ratio can be computed (1.7)

$$Æ := \frac{b^2}{S} \tag{1.7}$$

which is a measure of a wing's 'slenderness' (a large aspect ratio corresponds to a long, thin wing like on a glider; while a small $Æ$ corresponds to a stubby wing like on a fighter jet). Note that, for a rectangular wing with $S = bc$ for some cord c, this simplifies to $Æ = b/c$. Also, a note about language—for an aircraft, a 'wing' has two halves, one on each side of the aircraft; thus a typical jetliner has only one wing (unlike a bird, for which we typically talk of two wings). Think of the 'biplanes' and 'triplanes' used in World War I, which had two (respectively three) wings.

For a finite wing, we can approximate the wing's lift coefficient C_L from the sectional lift coefficient C_l as

$$C_L \approx \frac{Æ}{Æ + 2} C_l \tag{1.8}$$

which is fairly accurate for larger aspect ratios (see McCormick (1995) for more details, and more accurate (but more complex) models).

The additional drag experienced because of the finite wings is known as the *induced drag*, f_{D_i} (that is, the drag induced by the lift) and can be approximated as

$$f_{D_i} = \frac{1}{2} \rho v_\infty^2 S C_{D_i} \tag{1.9}$$

with the *induced drag coefficient* C_{D_i} predicted as a function of the wing lift coefficient and aspect ratio as

$$C_{D_i} \approx \frac{C_L^2}{\pi Æ} \tag{1.10}$$

In addition to the induced drag, a real aircraft will experience *parasite drag*. Predicting parasite drag is hard, and generally there is no way around empirical or computational results. The end is typically a model of the form

$$f_{D_0} = \frac{1}{2} \rho v_\infty^2 S C_{D_0} \tag{1.11}$$

and C_{D_0} found in a look-up table as function of (e.g.) Reynolds and Mach numbers.

An interesting consequence of this division of drag is that an aircraft at steady flight (constant speed and height) has a non-zero speed of minimum drag. The parasite drag (1.11) simply grows quadratically with speed, v_∞. However, as the speed increases, the lift coefficient required decreases, leading to a decrease in the induced drag coefficient (1.10). At low speeds, the total drag is dominated by induced drag, while at high speeds the parasite drag dominates. This is explored in Problem 1.7.

1.1.3 Aerodynamic Forces: Propeller Thrust

With some understanding of fixed wings, we now turn our attention to rotating wings, or propellers. A propeller produces a thrust force parallel to its axis of rotation. Such flows are inherently three-dimensional, and complex; moreover, a propeller typically is geometrically more complex than an aircraft wing, with the width, airfoil shape, and angle of attack all varying along the propeller from the hub to the tip. Here, we will use a 'blade element model' to model the propeller.

We may gain some insight though by performing a 'blade element' model of a propeller. Such a model is made by effectively chopping the propeller up into infinitesimally wide airfoil segments (of width dr), starting at the hub of the propeller $r = 0$ and moving radially outwards to the propeller tip $r = r_p$, the propeller radius, illustrated in Fig. 1.7. The blade element will create an infinitesimal lift force df_l perpendicular to the plane of rotation, and an infinitesimal drag force df_d opposing the rotation. We will do this specifically for the simpler case of a non-translating propeller, i.e. a propeller where the hub is not moving through space. These infinitesimal forces will have the usual aerodynamic forms, Eq. (1.5), with surface area $c(r)dr$ and lift coefficient $C_l(r)$:

$$df_l = \tfrac{1}{2}\rho v_\infty(r)^2 c(r) dr C_l(r) \tag{1.12}$$

Similarly, for the drag, we have

$$df_d = \tfrac{1}{2}\rho v_\infty(r)^2 c(r) dr C_d(r) \tag{1.13}$$

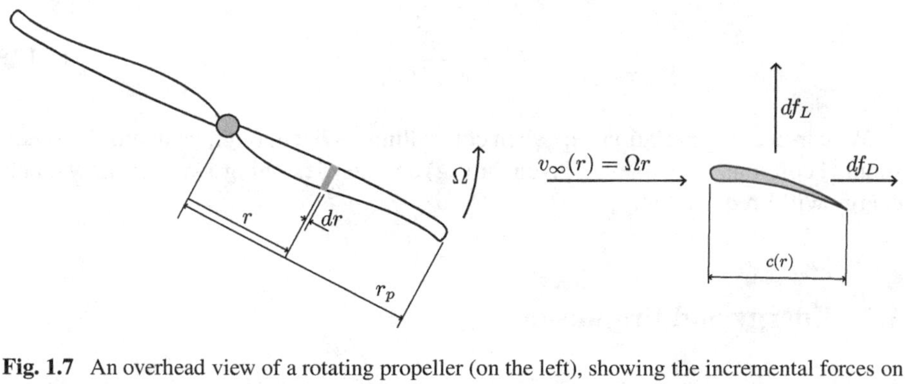

Fig. 1.7 An overhead view of a rotating propeller (on the left), showing the incremental forces on a blade element (right) of width dr located at a distance r from the propeller hub

Substituting the local flow $v_\infty(r) = \Omega r$, and then integrating the lift over $r \in [-r_p, r_p]$ yields that the thrust force f_T scales as

$$f_T = \int_{-R}^{R} d\, f_l = \int_{-R}^{R} \tfrac{1}{2}\, \rho v_\infty(r)^2 c(r) C_l(r) dr \tag{1.14a}$$

$$= \left(\int_{-R}^{R} \tfrac{1}{2} r^2 c(r) C_l(r) dr \right) \rho \Omega^2 \tag{1.14b}$$

$$=: c_T \rho \Omega^2 \tag{1.14c}$$

with c_T a propeller constant.

Noting that the drag force is an odd function of the radius, i.e. $df_d(-r) = -df_d(r)$, we get that there is zero net drag force, $f_d = 0$ (as expected, there's no net force in the plane of rotation). However, the torque n required to drive the propeller can also be computed by integrating the infinitesimal torque $dn = r df_d$, to get

$$n \propto \rho \Omega^2 \propto f_T \tag{1.15}$$

This is a surprisingly compact relationship—the torque is simply proportional to the force. For simplicity, introduce a force-to-torque constant γ so that $n = \gamma f_T$. We can compute the mechanical power at the propeller hub, P, from this as

$$P = n\Omega \propto f_T^{\frac{3}{2}} \tag{1.16}$$

In Sect. 8.2 we derive a relationship between propeller size (expressed as radius r_p), thrust f_T, power P, and air density ρ, repeated here for simplicity

$$P = \frac{f_T^{\frac{3}{2}}}{r_p \sqrt{2\pi\rho}} \tag{1.17}$$

We can use these relationship when controlling a system (e.g., what motor speeds should I command to achieve a given thrust?) or when designing a system (how much energy will I require to fly?).

1.2 Energy and Propulsion

All heavier than air flight requires the aircraft to do work to stay aloft. For the vehicles we consider, this typically is in the form of rotating a propeller to produce thrust, either for overcoming the drag while a wing provides lift; or driving the propellers that provide the vertical thrust.

1.2.1 Powertrain

An aircraft powertrain is the sequence of components required to provide primary power. Typically, this starts with an energy storage system, followed by potentially a sequence of conversion steps eventually ending up at a rotating shaft driving a propeller or turbine. We will consider two distinct propulsion models: combustion, and battery-electric. We note that many more options exist, for example fuel cells, or hybrid approaches combining batteries with combustion. There are also more exotic solutions, such as solar-powered systems (see, e.g., Nasa's Project Pathfinder[2]), or even systems powered by laser beams (Achtelik et al. 2011). The realities of the climate crisis make understanding powertrains even more important, beyond the (already complex and pressing) aspects of creating useful flight-capable systems.

A typical battery-electric powertrain thus consists of the following components: the *battery* converts stored chemical energy to electrical energy, which is then carried as direct current by wires to *motor controllers*. The motor controllers are often called Electronic Speed Controllers, or ESCs, and are responsible for converting the direct current power from the battery to phase-matched alternating current, provided to the *motors*. The electric motors are, for meaningfully sized systems, brushless electric motors. These motors then mechanically drive a shaft, with an attached *propeller*, which accelerates air to create thrust. Each component in the pipeline converts the power (e.g., from chemical to electric, from direct to alternating current, from electric to mechanical, and from mechanical to aerodynamic), and each component will suffer some inefficiencies in this conversion. The efficiency of any particular step is often a complex function of the system's operating conditions, for example the efficiency of a propeller varies significantly with aircraft flight conditions. Similarly, as a battery discharges, the voltage may drop, so that higher current is required for the same power, which leads to higher transmission losses in the electric system.

A combustion powertrain stores the energy in a (typically liquid) fuel, which is combusted in ambient air to generate heat. The heated, expanding gas is then harnessed to produce a torque, which is mechanically transmitted to a shaft.

As a first approximation, combustion fuels are capable of storing significantly more energy per unit mass (that is, they have a higher *specific energy*[3]). For larger systems where mass is at a premium, it is often very difficulty to move away from combustion fuel (for example, long-haul passenger aviation).

Combustion systems tend to have fewer motors than battery electric systems: because combustion engines rely on expanding gas to do work, they are more efficient the higher the internal temperatures are. In reality, some heat is lost through the engine, so that there is less capacity for the gas to do work. An efficient combustion engine thus strives to minimize the surface area that the hot gas can lose heat through, typically this means a single large engine is more efficient than many small engines. Moreover, once the chemical energy has been converted to mechanical power, it

---

[2] https://www.nasa.gov/centers/armstrong/news/FactSheets/FS-034-DFRC.html.

[3] Some people prefer the more confusing term 'gravimetric energy density', rather than 'specific energy', to describe a fuel's mass-normalized energy stored.

is difficult to transmit this mechanical power over long distances (drive shafts or hydraulic lines are heavy). We thus aim to locate the engine near the propeller.

For electric systems, these constraints are much less relevant. There are two reasons for this: first, electric power can be transmitted much more effectively through a structure than mechanical power (especially if high voltages are used). Next, the efficiency of electric motors does not scale as poorly as for combustion motors, so that an electric aircraft can more easily distribute the propulsion throughout the vehicle structure, with many motors driving many propellers. Such *distributed electric propulsion* may have multiple advantages, including allowing for greater safety through redundancy of actuators, or more maneuverability.

1.2.2 Power Consumption in Flight

Accurate prediction of power consumption in flight is extremely challenging, due, in part, to the variation of efficiency of components in the powertrain with operating conditions. However, we may build some intuition by looking only at the aerodynamic power, and assuming that the remainder of the powertrain has been designed efficiently.

Here we will provide an exemplary analysis for a fixed-wing aircraft, and a multi-rotor. For example, consider a multi-rotor aircraft with N_p propellers, each with radius r_p, and a total mass of m_Σ. The ideal aerodynamic power for propeller i, for such a vehicle in hover, is predicted by the actuator disk model (derived in Sect. 8.2) to be

$$P_i = \frac{f_i^{\frac{3}{2}}}{r_p\sqrt{2\pi\rho}} \tag{1.18}$$

with $f_i = \frac{1}{N_p} m_\Sigma g$. The total power consumed is thus

$$P_\Sigma = N_p P_i = \frac{(m_\Sigma g)^{\frac{3}{2}}}{r_p\sqrt{2\pi\rho N_p}} \tag{1.19}$$

We thus note that the power consumed scales super-linearly with the vehicle mass, and inversely proportionally to the square root of total propeller area, $\sqrt{N_p \pi r_p^2}$. This allows us to conclude that a lighter vehicle is more efficient (and possibly, significantly more efficient); but also that we can make a vehicle more efficient by increasing the total propeller area (either by using larger propellers, or by increasing the number of propellers). Of course, in practice, adding more propellers will increase the mass significantly, and increasing the propeller radius also requires heavier propellers, and may lead to other constraints being violated (e.g., the overall size of the vehicle may become too large, or the propeller tip speed may become too close to the speed of sound).

Note that a real propeller will require more power than the idealized model predicts, this is known as a propeller's *figure of merit*. This is defined as the ratio between the actual shaft power the propeller uses, and the aerodynamic power from (1.17). Notice that this is not the same thing as *efficiency*, which captures the ratio of output to input power of a device. For example, a hovering quadcopter has a stationary propeller, which produces thrust, but does no work (since it is stationary), so that the propeller has zero efficiency irrespective of how well/poorly it is designed (but its figure of merit is not zero).

For fixed-wing aircraft, we can use the *drag power* as a metric for efficiency (that is, the power associated with the drag force f_D applied at the aircraft speed v_∞). Noting that the total drag consists of induced drag f_{D_i} and parasite drag f_{D_0}. To a first approximation, for a given aircraft, we may model the parasite drag coefficient C_{D_0} as constant, and the induced drag coefficient C_{D_i} using the approximation (1.10). Since the parasite drag coefficient is constant, the power P_{D_0} associated with it will scale like the speed cubed:

$$P_{D_0} = f_{D_0} v = \frac{1}{2} \rho v^2 S C_{D_0} v \propto v^3 \tag{1.20}$$

For the induced drag, we note that the lift coefficient will decrease with increasing speed, so that the induced drag coefficient will decrease with increasing speed:

$$C_L = \frac{f_L}{\frac{1}{2} \rho v_\infty^2 S} \propto v^{-2} \tag{1.21}$$

and substituting into (1.10)

$$C_{D_i} \approx \frac{C_L^2}{\pi Æ} \propto v^{-4} \tag{1.22}$$

so that the induced drag force scales like $f_{D_i} \propto v^{-2}$, and the induced drag power as $P_{D_i} \propto v^{-1}$. Thus, at low air speeds, a fixed-wing aircraft will consume large amounts of power because of the induced drag (loosely, 'lift is expensive when flying slowly'). However, at high air speeds the aircraft also consumes large amounts of power, simply because of the parasite drag (even though the cost of lift becomes negligible as $v \to \infty$). There is thus an optimal speed where the sum of the parasite and induced drag is at a minimum, and the aircraft consumes least power. Note that this is known as the "maximum endurance" speed (for a fixed energy budget, the aircraft can maintain steady flight for the longest time at this speed). There is also the "maximum range" speed, which corresponds to the speed at which the force is minimized (for a fixed energy budget, the aircraft can cover the longest distance at this speed). This is explored in Problem 1.7.

1.3 Actuators and Control Surfaces

We modify the flight of an aircraft through actuators and control surfaces. One actuator that we've already discussed is a motor driving a propeller, that can vary the thrust force produced by varying the speed of the propeller. There are, however, many additional actuators on typical aerial systems, and we describe a few here.

A typical fixed-wing aircraft has four independent inputs, illustrated in Fig. 1.8. The first is the thrust magnitude, typically controlled by a varying a propeller speed, and/or varying the pitch of the propeller; we've discussed how to model thrust in Sect. 1.1.3. On each side of the main wing, the *aileron* is used to roll the aircraft around an axis passing through its nose. The ailerons actuate anti-symmetrically (that is, the left wing's will deflect upwards if the right wing deflects downwards). By deflecting, the aileron changes the shape of the airfoil of the wing, thereby changing the corresponding lift for this section of the wing; the anti-symmetric actuation thus causes an increase in lift on one side, with decrease on the opposite, thereby yielding moment.

The aircraft tailplane has two actuators. A moment to pitch the nose of the aircraft up or down is created by deflecting the *elevator*, located on the horizontal tailplane. Note that the elevator is actuated symmetrically. Lastly, the *rudder* is located on the vertical tailplane, and acts to create a yawing moment to turn the aircraft nose left/right. Of course, in practice, each aerodynamic surface will cause changes in local drag as well as lift, and their effects are typically more complex than 'ailerons cause only roll moments'. However, to a first approximation we may imagine that the aircraft has an aerodynamic actuator for each component of total moment.

For smaller craft, the aerodynamic surfaces may be driven by directly linked, electrically-driven actuators (e.g., a servo motor), while large craft typically use hydraulic actuation. Significant thought must be given to safety and redundancy (what happens if an actuator becomes 'stuck', or experiences a 'hardover' and goes uncommanded to maximum deflection, or loses all actuation and becomes 'floating').

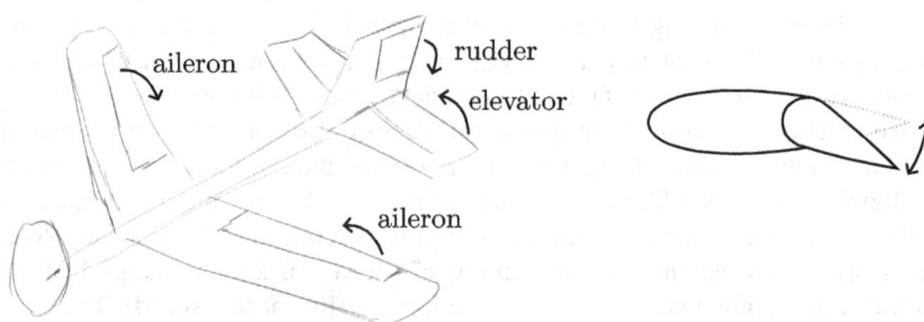

Fig. 1.8 Control surfaces on a typical fixed-wing aircraft, on the right; with a cross-section of a wing showing how the aileron changes the airfoil shape

Problems for Chapter 1

Problem 1.1 (*10 points*) Consider a typical light aircraft, with mass 900 kg. The aircraft has a 16 m^2 wing, with maximum lift coefficient $C_{L,\max} = 1.6$. What is the minimum speed $v_{\min}$ that the aircraft can fly at without losing altitude (this is also the "stall speed" of the aircraft)?

Use $\rho = 1.225\,\text{kg/m}^3$.

Problem 1.2 (*10 points*) In this problem, we will calculate the flow regime for various aircraft. For your own sanity, complete this using a spreadsheet. Use atmospheric data as provided in the Appendix A. We will look at the following aircraft, operating in steady state at the given conditions:

a. The *Airbus A380*, with wing planform information as given in en.wikipedia.org/wiki/Airbus_A380, operating at 11 km above sea level at a speed of 903 km/h, with mass 500,000 kg.
b. The *DG Flugzeugbau DG-808C* glider plane, with data available at en.wikipedia.org/wiki/DG_Flugzeugbau_DG-800, with operating altitude at 5 km above sea level at speed 150 km/h with mass 600 kg.
c. The *Lockheed Martin F-22* fighter jet, with data available at en.wikipedia.org/wiki/Lockheed_Martin_F-22_Raptor, operating at 11 km above sea level at speed 1900 km/h with mass 29,410 kg.
d. The *Sensefly eBee* mapping UAV, with data available at www.sensefly.com/drone/ebee-mapping-drone/ (note—not all info is listed, and the wing area can be estimated from the photograph as around 0.44 m^2), operating at sea level at 60 km/h, with mass 0.7 kg.

Create a table, with one column for each of the aircraft, with the following rows (all values at the given operating condition):

a. Reynolds number (for the characteristic length in the Reynolds number, use the wingspan b)
b. Mach number
c. Wing aspect ratio $\mathcal{R}$
d. Lift coefficient
e. Induced drag force (estimated using the formula provided (1.10)).

Consider the different purposes of the aircraft—do the numbers make sense in their different contexts?

Problem 1.3 (*10 points*) You are designing a quadcopter, and want to estimate the power required. The vehicle weighs 0.5 kg, and has *four* propellers with thrust/torque characteristics as follows:

$$f_T = C_T \Omega^2$$
$$\tau = \gamma f_T$$

with Ω the individual propeller speed, f_T (γ) the force (torque) per propeller, and

$$C_T = 6.41 \times 10^{-6} \mathrm{N}/(\mathrm{rad/s})^2$$
$$\gamma = 0.017\,\mathrm{m}.$$

(a) Compute the propeller speeds at hover (assume each propeller carries an equal weight).
(b) Compute the total mechanical power required at hover.

Problem 1.4 (*10 points*) Consider the following wing design, with all dimensions in [mm]:

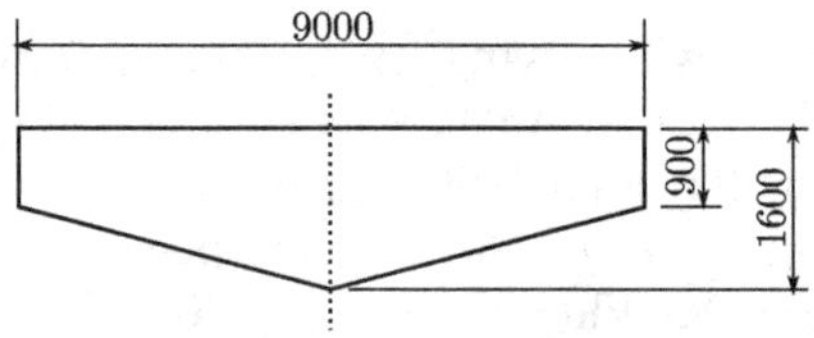

The wing is made from a flat plate, with lift coefficient dependent on angle of attack α as $C_l = 2\pi\alpha$. The wing is attached to an aircraft with total mass 1000 kg. You may neglect any aerodynamic effects of the aircraft body, tail, etc. Assume operation at sea level in the standard atmosphere.

(a) Compute the wing's aspect ratio ($\mathcal{R}$).
(b) Estimate the wing's lift coefficient C_L as a function of the wing's angle of attack. (Note: the reference area for the wing coefficients is the wing surface area when seen from the top).
(c) At a cruise speed of $55\,\mathrm{m\,s^{-1}}$, estimate the required angle of attack for level flight.

Problem 1.5 (*12 points*) Estimate (to within $\pm10\%$) the velocity with which a bowling ball dropped from a great height will hit the ground (that is, its terminal velocity). You may model the bowling ball as a smooth sphere, of diameter 0.217 m and with mass 7.0 kg. The drag coefficient can be read off from the following figure (where the diameter is used as the Reynolds number reference length, and the reference area is the sphere's projected surface area in the direction of the velocity, with the figure generated from data in the Python Fluids package (Bell and Contributors 2025):

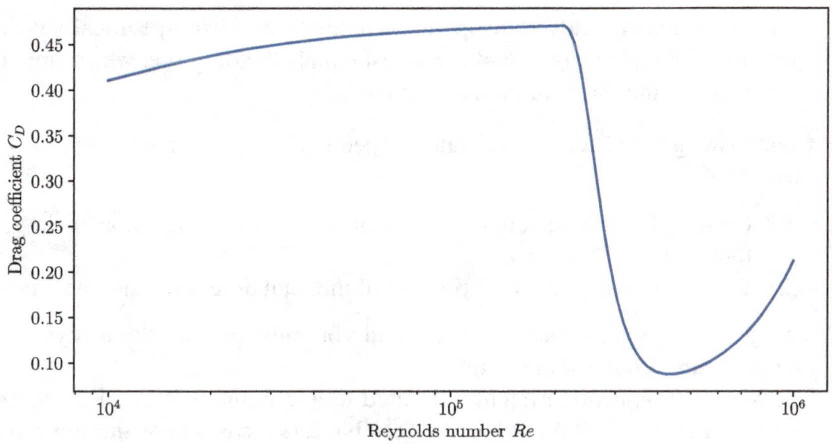

Model the atmosphere as having constant density (ρ) and viscosity (μ), with specifically $\rho = 1.225\,\text{kg/m}^3$ and $\mu = 18.27 \times 10^{-6}\,\text{Pa s}$.

Ungraded aside: notice that, near $Re \approx 3 \times 10^5$, the drag coefficient decreases with increasing Reynolds number (that is, with increasing speed). Does this mean that a falling sphere at that Reynolds number will accelerate, because increasing the speed decreases the drag (and the faster it falls, the faster it 'wants to fall')?

Problem 1.6 (*7 points*) You are working with a fixed-wing UAV. The vehicle has a total mass of 1 kg, and has a rectangular square wing with span $b = 2\,\text{m}$ and chord $c = 0.2\,\text{m}$, where the wing is a flat plate. You want the vehicle to fly at a constant speed of 20m/s in an atmosphere with density $\rho = 1.2\,\text{kg/m}^3$.

Estimate the required angle of attack for the wing.

Problem 1.7 (*13 points*) In this problem we investigate the feasible range of flight speeds for a fixed-wing aircraft. Specifically, we want to compute the total drag at various airspeeds, for the fixed-wing UAV shown below.

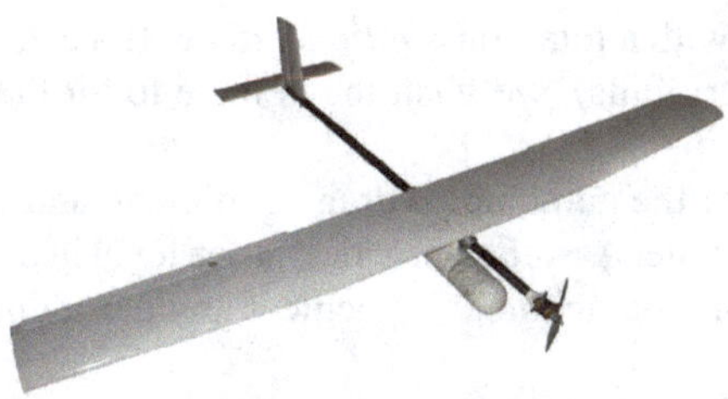

The coefficient for the parasite drag is $C_{D_p} = 0.0356$ (with reference area the wing surface area $S = 0.69\,\text{m}^2$), and the wing aspect ratio is $\mathit{AR} = 8.75$. The total mass is $m = 5.3\,\text{kg}$.

We are interested in steady-state operation of the UAV, where specifically the total lift equals the vehicle's weight, and the thrust equals the drag (i.e. we assume that, as airspeed varies, the thrust remains horizontal).

(a) Create two graphs, for steady-state airspeeds v_∞ in the range between 3 and 30 m/s, of:

    (i) the drag force, with a line for each of the induced drag, parasite drag, and total drag.
    (ii) the drag power, that is the product of the total drag force and the airspeed.

    The graphs must be clearly labelled, and you must provide the analysis used, and the source code for the graphs.

(b) The aircraft propulsion system is limited to a maximum force of 10 N, and a maximum power 200 W. What are feasible airspeeds where the aircraft can maintain steady flight? Assume that there are no losses in the drivetrain.

(c) We define (i) the vehicle's *maximum range speed* v_r^* as that airspeed at which it can cover the largest distance for a given amount of fuel, i.e. where the total drag *force* is minimized, and (ii) the vehicle's *maximum endurance speed* v_e^* as that airspeed at which it can stay aloft for the longest amount of time, i.e. where the drag *power* is minimized. What are these two optimal airspeeds for this vehicle? Also mark them on your graph of (a).

Problem 1.8 (*5 points*) You are working on a high performance quadcopter. The quadcopter has a total mass of 1 kg, and has four propellers. During high-speed flight, the quadcopter experiences a drag force—you may model the aerodynamic drag of the quadcopter as that of a sphere with a projected surface area of $0.2\,\text{m}^2$, with a constant drag coefficient of 0.5 (both independent of the quadcopter's orientation).

You wish to design a quadcopter that can climb vertically at 20 m/s. Assume that the density of air is $1\,\text{kg/m}^3$.

How much thrust does each propeller need to be able to produce for the quadcopter to achieve this? You may assume that the propeller thrust is unaffected by the vehicle motion.

Problem 1.9 (*4 points*) You are tasked with designing an emergency parachute system for a quadcopter with a total mass $m^B = 2.5\,\text{kg}$. If a catastrophic system failure occurs (e.g., the battery fails), we wish the vehicle to hit the ground at a speed no greater than $v_{\text{max}} = 1\,\text{m/s}$.

Take $C_D = 1.75$ for the parachute's drag coefficient, and determine the diameter of the necessary parachute. Assume operation at sea level in the standard atmosphere. Neglect the mass of the parachute, and neglect any aerodynamic forces beside the drag of the parachute.

Problem 1.10 (*10 points*) You are working on a fixed-wing aircraft. The aircraft has a rectangular wing, with span 12 m and width 2 m. The wing is made of an airfoil, with the below sectional lift characteristics C_l, as plotted against angle of attack α. The aircraft has a total mass of 700 kg.

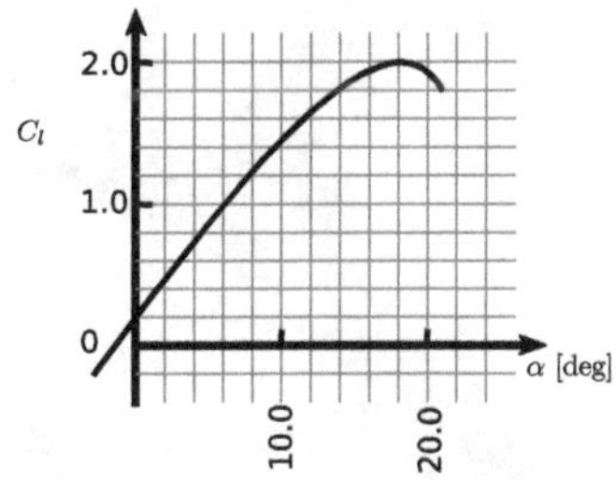

Estimate the lowest speed at which the aircraft can maintain steady flight at sea level in the standard atmosphere.

Problem 1.11 (*10 points*) A Boeing 787 has a wingspan of 60 m, a total wing surface area of 377 m^2, and a total mass of 250, 000 kg.

Approximating the wing as a flat plate, calculate the angle of attack necessary to take off at sea level, in the standard atmosphere. The speed at take off is 85 m s^{-1}.

Note: Neglect all forces aside from weight, and the lift of the main wing.

References

Achtelik MC, Stumpf J, Gurdan D, Doth K-M (2011) Design of a flexible high performance quadcopter platform breaking the MAV endurance record with laser power beaming. In: 2011 IEEE/RSJ international conference on intelligent robots and systems. IEEE, pp 5166–5172

Anderson J (2011) Fundamentals of aerodynamics (SI units). McGraw Hill

Bell C (2025) Contributors, fluids: fluid dynamics component of chemical engineering design library (ChEDL). https://github.com/CalebBell/fluids

McCormick BW (1995) Aerodynamics aeronautics and flight mechanics. Wiley

Chapter 2
Three-Dimensional Geometry

Without geometry, life is pointless.

In this chapter we look at how to describe the motion of a body in 3D space, with a focus on displacements, and rotations. We strive to introduce concepts 'intuitively', that is utilizing everyday concepts (after all, we live in a 3D world); this allows us to make the connection between the 3D world around us, and the linear algebra with which we can encode things. We follow the notation and style of Zipfel (2007), and the mantra therein: "from tensor modelling to matrix coding".

The specific objectives of this chapter are to develop a firm grasp of how to describe relative position and orientation of objects in 3D space. We will do this using the language of *displacement* of points, and *rotation* of frames. The link between linear algebra in $\mathbb{R}^3$ and the 3D world will be crucial.

2.1 Nomenclature

First, we describe our notation and symbols. The style may, at first, appear tedious, but it is well worth the effort to create a clear language that is unambiguous.

2.1.1 Invariant Description

A key concept is to describe the system of interest in a *coordinate-free* fashion, that is the idea that the fundamental nature of the system does not depend on the coordinate

M. W. Mueller, *Dynamics and Control of Autonomous Flight*, Mathematical Engineering,
https://doi.org/10.1007/978-3-031-91871-1_2

system being used,[1] i.e. nature is independent of our description thereof. The way we will get there is to argue that our description holds with any coordinate system, in other words it is independent of the coordinate system chosen.

We will use the following notational convention. Although we will define many terms only later, we summarize them here for reference.

- We will use specific letters to indicate the type of a quantity of interest (e.g., an s consistently means 'displacement', and R a 'rotation').
- An underbar will indicate a tensor quantity, with lower-case being first order tensors like $\underline{v}$ and upper-case being second-order tensors like $\underline{V}$ (we'll define 'tensors' in a moment).
- A scalar will be written as simply a letter, like v.
- A tensor superscript will refer to a *frame* or a body (with an associated frame), e.g. $\underline{v}^A$, and the frame will always be capitalized.
- A tensor subscript will refer to a *point*, e.g. $\underline{v_A}$, and the point will always be capitalized.
- $\frac{\mathrm{d}}{\mathrm{d}t}(\cdot)$ denotes the ordinary time derivative.
- $D^A\underline{B}$ refers to the rotational time derivative of tensor quantity $\underline{B}$ with respect to frame A (again, defined later).
- A bar over a tensor is used to indicate its *transpose*, e.g. $\overline{\underline{x}}$.
- Square braces, with a superscript, will indicate the expression of a tensor in a coordinate system, e.g. $\left[\,\underline{x}\,\right]^B$. This is a vector/matrix, and behaves exactly as any other matrix you are used to from linear algebra.

Below are some special tensors, that will appear often (and are defined later):

$\underline{s_{AB}}$	the displacement of point A with respect to point B.
$\underline{v_B^I}$	the velocity of point B with respect to the frame I.
$\underline{\omega^{BI}}$	the angular velocity of frame B with respect to frame I.
$\underline{J_C^B}$	the inertia tensor of body (frame) B with reference point C.
$\underline{I}$	the identity tensor.

2.1.2 *Coordination and Coordinate Systems*

In this subsection, we will start with a hands-on definition of tensors and coordinates, before formalizing various definitions.

Consider two points in the world, A and B (for example, the point A is the bottom-right corner of this page, and B is the page number on this page). We can draw an arrow on the page, showing the displacement of point A with respect to point B, $\underline{s_{AB}}$ (this is an arrow starting in the page number, and ending at the bottom-right page corner). We call this a *displacement tensor*, and we note that we can measure

[1] There are some philosophical issues here too, see Zipfel (2007, Sect. 2.1.3). Note also that this is different than the "coordinate-free" that is used in *geometric approaches* to dynamics, such as Lee et al. (2017).

the size of this displacement by laying a ruler along the displacement; we will write the measured size $\| \underline{s_{AB}} \|$.

Given two other points, C and D, we can define another displacement tensor $\underline{s_{CD}}$, and measure its length as $\| \underline{s_{CD}} \|$. We now define a *scalar product* of two tensors using the notation $\sigma(\underline{s_{AB}}, \underline{s_{CD}})$ (think of it as the "σcalar product"), which requires us measuring the angle between the tensors $\angle(\underline{s_{AB}}, \underline{s_{CD}})$, so that

$$\sigma(\underline{a}, \underline{b}) := \| \underline{a} \| \, \| \underline{b} \| \cos(\angle(\underline{a}, \underline{b})) \tag{2.1}$$

You may know this operation as the 'dot product', though this name is unfortunate.[2] Note that this computation requires only a ruler and protractor. We call tensors *orthogonal* if their scalar product is zero (i.e., the angle is 90°).

Clearly, from the definition, $\sigma(\underline{a}, \underline{b}) = \sigma(\underline{b}, \underline{a})$. Furthermore, in Problem 2.20 we prove that this operator is linear, i.e., for some first-order tensors $\underline{a}$, $\underline{b}$, and $\underline{c}$; and scalar α, we have:

$$\sigma(\underline{a}, \alpha\underline{b}) = \alpha\sigma(\underline{a}, \underline{b}) \tag{2.2}$$

$$\sigma(\underline{a}, \underline{b} + \underline{c}) = \sigma(\underline{a}, \underline{b}) + \sigma(\underline{a}, \underline{c}) \tag{2.3}$$

Next, we define a coordinate system, say the paper-fixed P system, by defining a series of three first-order tensors (which we'll call $\underline{1^P}$, $\underline{2^P}$, and $\underline{3^P}$) that satisfy:

1. Each component has length one: $\| \underline{1^P} \| = \| \underline{2^P} \| = \| \underline{3^P} \| = 1$.
2. All components are orthogonal: $\sigma(\underline{1^P}, \underline{2^P}) = \sigma(\underline{1^P}, \underline{3^P}) = \sigma(\underline{2^P}, \underline{3^P}) = 0$.
3. The three components follow the "right-hand rule", i.e., if you take your right hand, point your index finger along $\underline{1^P}$, your middle finger along $\underline{2^P}$, then your thumb extends along $\underline{3^P}$.

We can take any first-order tensor $\underline{v}$ and break it into three unique *coordinates*, so that

$$\underline{v} = v_1 \underline{1^P} + v_2 \underline{2^P} + v_3 \underline{3^P} \tag{2.4}$$

where we can compute the coordinates with the scalar product (e.g., $v_1 = \sigma(\underline{v}, \underline{1^P})$), and the summation means taking the step of size v_1 along the direction $\underline{1^P}$, followed by a step of size v_2 in direction $\underline{2^P}$, etc. We call the three scalars v_1, v_2, v_3 the coordinates of $\underline{v}$ in the coordinate system P, which we'll write by enclosing the tensor in square brackets, with the superscript indicating the frame, as

$$[\underline{v}]^P = \begin{bmatrix} v_1 \\ v_2 \\ v_3 \end{bmatrix} \in \mathbb{R}^3 \tag{2.5}$$

[2] There is no information in the name *dot product of two vectors* (the dot is an arbitrary choice of multiplication symbol), while *scalar product of two vectors* clearly describes the result.

Thus, a coordinate system allows us to take a physical object in the world (e.g., the displacement between two points) and transform it into a linear algebra object (a 3-vector). We'll sometimes save space, and write a column vector as $\left[\underline{v}\right]^P = (v_1, v_2, v_3)$.

We can introduce a second coordinate system, say the D system fixed to the desk, with corresponding directions $\underline{1^D}$, $\underline{2^D}$, and $\underline{3^D}$. Consider now the expression of the vector $\underline{v}$ in the page system P as $\left[\underline{v}\right]^P = (v_1^P, v_2^P, v_3^P)$, and in the desk system D as $\left[\underline{v}\right]^D = (v_1^D, v_2^D, v_3^D)$.

We note that we can always write the base tensors of the D system using the P system, which will require nine scalars $t_{i,j}$:

$$\underline{1^P} = t_{1,1}\,\underline{1^D} + t_{1,2}\,\underline{2^D} + t_{1,3}\,\underline{3^D} \tag{2.6}$$
$$\underline{2^P} = t_{2,1}\,\underline{1^D} + t_{2,2}\,\underline{2^D} + t_{2,3}\,\underline{3^D} \tag{2.7}$$
$$\underline{3^P} = t_{3,1}\,\underline{1^D} + t_{3,2}\,\underline{2^D} + t_{3,3}\,\underline{3^D} \tag{2.8}$$

or, more compactly, as

$$\begin{bmatrix} \underline{1^P} \\ \underline{2^P} \\ \underline{3^P} \end{bmatrix} = \begin{bmatrix} t_{1,1} & t_{1,2} & t_{1,3} \\ t_{2,1} & t_{2,2} & t_{2,3} \\ t_{3,1} & t_{3,2} & t_{3,3} \end{bmatrix} \begin{bmatrix} \underline{1^D} \\ \underline{2^D} \\ \underline{3^D} \end{bmatrix} \tag{2.9a}$$

$$=: M \begin{bmatrix} \underline{1^D} \\ \underline{2^D} \\ \underline{3^D} \end{bmatrix} \tag{2.9b}$$

with $M \in \mathbb{R}^{3\times 3}$ is a matrix that we determined only from scalar products of the tensors defining the two systems, and

$$t_{i,j} = \sigma\left(\underline{i}^P, \underline{j}^D\right) = \cos\left(\angle\left(\underline{i}^P, \underline{j}^D\right)\right). \tag{2.10}$$

We can now investigate how the coordinates in the P system, $\left[\underline{v}\right]^P$, relate to those in the D system, $\left[\underline{v}\right]^D$:

$$\underline{v} = \underline{v} \tag{2.11}$$
$$v_1^D\,\underline{1^D} + v_2^D\,\underline{2^D} + v_3^D\,\underline{3^D} = v_1^P\,\underline{1^P} + v_2^P\,\underline{2^P} + v_3^P\,\underline{3^P} \tag{2.12}$$

Factoring gives

$$\begin{bmatrix}\underline{1}^D & \underline{2}^D & \underline{3}^D\end{bmatrix}\begin{bmatrix}v_1^D\\ v_2^D\\ v_3^D\end{bmatrix} = \begin{bmatrix}\underline{1}^P & \underline{2}^P & \underline{3}^P\end{bmatrix}\begin{bmatrix}v_1^P\\ v_2^P\\ v_3^P\end{bmatrix} \tag{2.13a}$$

$$= \begin{bmatrix}\underline{1}^D & \underline{2}^D & \underline{3}^D\end{bmatrix} M^T \begin{bmatrix}v_1^P\\ v_2^P\\ v_3^P\end{bmatrix} \tag{2.13b}$$

where we used the matrix transpose M^T.

Comparing the above, we can conclude that for any $\underline{v}$, we have

$$\left[\underline{v}\right]^D = M^T \left[\underline{v}\right]^P \tag{2.14}$$

or that the act of expressing a first-order tensor in a different coordinate system is equivalent to the linear algebra operation of a matrix multiplication.

Understanding this matrix is central to this course, and we will call it a *transformation matrix*.[3] Note that this is a purely descriptive change (and does not change the underlying physical quantity, e.g., the displacements of the points in the world is unchanged by which coordinate systems we may choose to use). For example, changing to coordinate system B from system A is achieved by multiplication with the transformation $[T]^{BA} \in \mathbb{R}^{3\times 3}$:

$$\left[\underline{x}\right]^B = [T]^{BA} \left[\underline{x}\right]^A \tag{2.15}$$

We read this as $[T]^{BA}$ is the transformation matrix of coordinate system B with respect to coordinate system A.

There is a subtle distinction between a *frame* and a *coordinate system*: a frame is a physical object, which is made up of a set of points which are fixed with respect to one another. A coordinate system is a mathematical tool, used to coordinate tensors (i.e. write them as numbers), and a change of coordinate system is a change in description (but the world is the same as before). A change of frame, however, involves changing the world (e.g. rotating a body). Thus, when we are modelling, we will use frames to describe the world; when we are encoding our models, we will use coordinate systems. Of course, we will often attach our coordinate systems to a particular frame, and the terms may then be interchanged without confusion (e.g., the body-fixed frame defines the body-fixed coordinate system).

Another important fact that may be distinct from your prior experience in geometry, or other definitions of coordinate systems: for us, a coordinate system is a set of directions, but there are no *preferred points*, and specifically a coordinate system does not have an *origin* (remember, points are physical objects in the world, while coordinate systems are abstract constructions).

[3] This is also sometimes called the 'Direction Cosine Matrix' (DCM), a name that follows from (2.10).

Example 2.1 Consider the following example. We will look at a world with three points, A, B, and C; the point A is a distance of 1 unit from both B and C, and the distance between B and C is $\sqrt{2}$. We also define six directions, from which we will have 2 coordinate systems. The directions are $\underline{1}^P$, $\underline{2}^P$, $\underline{3}^P$, $\underline{1}^Q$, $\underline{2}^Q$, and $\underline{3}^Q$. Note that $\underline{3}^P$ and $\underline{3}^Q$ are identical, and simply point out of the page. The direction $\underline{1}^Q$ is rotated by 45° with respect to $\underline{1}^P$ (similarly $\underline{2}^Q$ with respect to $\underline{2}^P$); and $\underline{1}^P$ is perpendicular to both $\underline{2}^P$ and $\underline{3}^P$.

The coordinate system P is defined using the three directions $\underline{1}^P$, $\underline{2}^P$, $\underline{3}^P$; similarly the system Q.

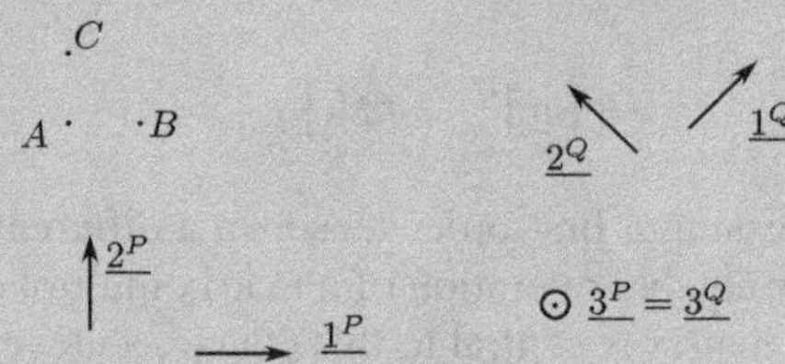

We can now express the displacement of point B with respect to point A by measuring the distance of B from A along the three directions P. Similarly, we can coordinate the remaining displacements, in both P and Q to get:

$$\left[\underline{s_{BA}}\right]^P = (1, 0, 0)\,, \qquad \left[\underline{s_{CA}}\right]^P = (0, 1, 0)\,, \qquad \left[\underline{s_{CB}}\right]^P = (-1, 1, 0)\,,$$
$$\left[\underline{s_{BA}}\right]^Q = \left(\tfrac{\sqrt{2}}{2}, -\tfrac{\sqrt{2}}{2}, 0\right), \quad \left[\underline{s_{CA}}\right]^Q = \left(\tfrac{\sqrt{2}}{2}, \tfrac{\sqrt{2}}{2}, 0\right), \quad \left[\underline{s_{CB}}\right]^Q = \left(0, \sqrt{2}, 0\right)$$

Consider now the below matrix, and convince yourself that $\left[\underline{s_{BA}}\right]^Q = [T]^{QP}\left[\underline{s_{BA}}\right]^P$

$$[T]^{QP} = \begin{bmatrix} \frac{\sqrt{2}}{2} & \frac{\sqrt{2}}{2} & 0 \\ -\frac{\sqrt{2}}{2} & \frac{\sqrt{2}}{2} & 0 \\ 0 & 0 & 1 \end{bmatrix}$$

and similarly for the other displacements given above.

Coordinating Equations

Once we have coordinated the tensor equations, we are left with a matrix equation, which behaves according to the usual rules of matrices and vectors. Coordinating a tensor equation is achieved by coordinating each component individually, for example, given the tensor equation $\underline{a} = \underline{b} + \underline{\underline{C}}\,\underline{d}$ and a coordinate system W, we coordinate the equation to get

$$\left[\underline{a}\right]^W = \left[\underline{b}\right]^W + \left[\underline{C}\right]^W \left[\underline{d}\right]^W \tag{2.16}$$

This ability to express in *any coordinate* system is, of course, the key idea of tensors, referring back to Sect. 2.1.1.

Identity Tensor

The mutliplicative identity tensor $\underline{I}$ satisfies, for any other tensor $\underline{a}$,

$$\underline{a} = \underline{I}\,\underline{a} \tag{2.17}$$

from which follows that the identity tensor is the same in any coordinate system C, and is the matrix identity:

$$\left[\underline{I}\right]^C = \begin{bmatrix} 1 & 0 & 0 \\ 0 & 1 & 0 \\ 0 & 0 & 1 \end{bmatrix}. \tag{2.18}$$

2.2 Tensors

2.2.1 Properties of Tensors

Let A and B be any allowable coordinate systems. Then the following defines Zipfel (2007) first- and second-order tensors:

Definition 2.1 A first-order tensor $\underline{x}$ is the collection of 3-tuples that satisfy the transformation $\left[\underline{x}\right]^B = [T]^{BA}\left[\underline{x}\right]^A$.

Definition 2.2 A second-order tensor $\underline{X}$ is the collection of 9-tuples that satisfy the transformation $\left[\underline{X}\right]^B = [T]^{BA}\left[\underline{X}\right]^A \overline{[T]}^{BA}$.

Because this has to hold for all coordinate systems A and B, the tensors describe something independent of the description we choose (and thus work for our requirement that the world does not depend on our description of it).

We may extend this to talk about *zeroth-order tensors*, which are simply scalars. These are objects that have no directionality (e.g. mass), and can thus be written down unambiguously without any coordinate system.

Example 2.2 Consider three points, fixed in the world: A, B, and C. Let $\underline{s_{AB}}$ be the displacement of A from B, $\underline{s_{AC}}$ be the displacement of A from C, and so on.

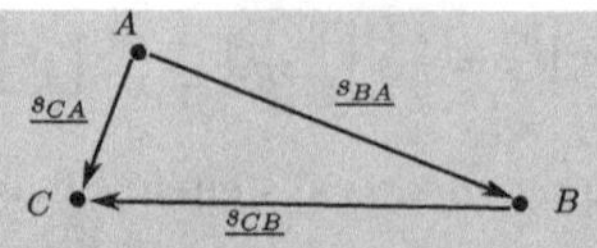

We note a few facts about our notational convention:

- $\underline{s_{CB}} + \underline{s_{BA}} = \underline{s_{CA}}$: adding vectors is equivalent to contracting the subscripts.
- $\underline{s_{BA}} + \underline{s_{AB}} = \underline{0}$, or $\underline{s_{BA}} = -\underline{s_{AB}}$: changing the subscript order is equivalent to flipping the sign.

2.2.2 *Algebra of Tensors*

The intuition you have from linear algebra is really all you need here: first-order tensors operate like vectors and second-order tensors like matrices. Nonetheless, we will describe some of the operations that tensors allow.

Two Common Operations: Transpose and Skew-Symmetric Operator

Before we get to the algebra, we define two common operations we may perform on tensors.

We first define the *transpose* of a tensor (denoted with an overbar) as $\overline{\underline{x}}$ as follows. Let $\underline{x}$ be a first order tensor, and then the transpose (written as $\overline{\underline{x}}$) is defined as below. For any coordinate system C, where we denote $\left[\underline{x}\right]^C = (x_1, x_2, x_3) \in \mathbb{R}^3$:

$$\left[\overline{\underline{x}}\right]^C = \overline{\left[\underline{x}\right]^C} = \left[x_1\ x_2\ x_3\right] \in \mathbb{R}^{1\times 3} \tag{2.19}$$

that is, the coordination of a transposed tensor is the transpose of the coordinated tensor. This can also be done for a second order tensor as well.

Secondly, we define the *skew symmetric operator*, written as $\underline{K}(\underline{x})$ as below, where again $\left[\underline{x}\right]^C = (x_1, x_2, x_3)$ is a first order tensor coordinated in an arbitrary coordinate system C:

$$\left[\underline{K}(\underline{x})\right]^C = \begin{bmatrix} 0 & -x_3 & x_2 \\ x_3 & 0 & -x_1 \\ -x_2 & x_1 & 0 \end{bmatrix} \tag{2.20}$$

or, written without coordinate systems, we define a new second order tensor $\underline{X}$ from the first order tensor $\underline{x}$ as $\underline{X} = \underline{K}(\underline{x})$. The proof that this is, in fact, a second-order tensor is left to Problem 2.18.

Addition of First-Order Tensors

Specifically, first-order tensors can be added, and obey the commutative and associative rules:

$$\underline{s_{CB}} + \underline{s_{BA}} = \underline{s_{BA}} + \underline{s_{CB}} \tag{2.21}$$

$$\left(\underline{s_{CA}} + \underline{s_{CB}}\right) + \underline{s_{BA}} = \underline{s_{CA}} + \left(\underline{s_{CB}} + \underline{s_{BA}}\right) \tag{2.22}$$

Multiplication of First-Order Tensors by Scalars

We can multiply tensors by scalars, and this multiplication is commutative, associative and distributive. For scalars a, b

$$a\left(b\,\underline{s_{CB}}\right) = b\left(a\,\underline{s_{CB}}\right) = (ab)\,\underline{s_{CB}} \tag{2.23}$$

$$(a+b)\,\underline{s_{CB}} = a\,\underline{s_{CB}} + b\,\underline{s_{CB}} \tag{2.24}$$

Multiplication of First-Order Tensors with First-Order Tensors

We can also multiply first-order tensors, and there are three multiplications of interest, scalar, vector, and dyadic products.

The *scalar product*, or inner product, of two tensors $\underline{a}$ and $\underline{b}$ is given by $\overline{\underline{a}}\,\underline{b}$, and returns a scalar. Earlier, in (2.1), we used the notation $\sigma(\underline{a}, \underline{b})$ to indicate this, and defined it using lengths and angles. We will now use this more compact notation, noting that the two operations are equivalent[4]; that is for any $\underline{a}$ and $\underline{b}$, we have

$$\sigma(\underline{a}, \underline{b}) = \overline{\underline{a}}\,\underline{b} \tag{2.25}$$

When coordinated in a coordinate system C, with $\left[\underline{a}\right]^C = (a_1, a_2, a_3)$ and $\left[\underline{b}\right]^C = (b_1, b_2, b_3)$ we have

$$\overline{\left[\underline{a}\right]}^C \left[\underline{b}\right]^C = \begin{bmatrix} a_1 & a_2 & a_3 \end{bmatrix} \begin{bmatrix} b_1 \\ b_2 \\ b_3 \end{bmatrix} = a_1 b_1 + a_2 b_2 + a_3 b_3 \tag{2.26}$$

Note that both the linear algebra, and the ruler & protractor definition imply

$$\overline{\underline{a}}\,\underline{b} = \overline{\underline{b}}\,\underline{a} \tag{2.27}$$

$$\overline{\underline{a}}\,\underline{a} = \left\| \underline{a} \right\|^2 \tag{2.28}$$

[4] Can you convince yourself that the ruler & protractor definition of $\sigma(\cdot, \cdot)$ in (2.1) yields the same result as the linear algebra operation? Do this by fixing a coordinate system C, and comparing $\sigma\left(\left[\underline{a}\right]^C, \left[\underline{b}\right]^C\right)$ and $\overline{\left[\underline{a}\right]}^C \left[\underline{b}\right]^C$.

where $\|\cdot\|$ is the Euclidean norm. Furthermore, since the result is scalar, we can omit the coordinate system use, and write simply $\overline{[\underline{a}]}^B [\underline{b}]^B = \overline{\underline{a}}\,\underline{b}$.

The second is the *vector product* of $\underline{a}$ and $\underline{b}$, which is often called the cross product.[5] For this, we use the skew-symmetric form of the first-order tensor $\underline{a}$. We can define a new first-order tensor $\underline{z}$ with the vector product as $\underline{z} = \underline{K}(\underline{a})\,\underline{b}$. This is equivalent to the notation that you may have seen elsewhere for the cross product, that $\underline{z} = \underline{a} \times \underline{b}$.

Note that our notation is vastly superior to the 'cross product' notation. Tensor multiplication is associative, so that $\underline{K}(\underline{a})\,\underline{K}(\underline{b})\,\underline{c}$ is unambiguous, since $(\underline{K}(\underline{a})\,\underline{K}(\underline{b}))\,\underline{c} = \underline{K}(\underline{a})\,(\underline{K}(\underline{b})\,\underline{c})$. The same does not hold for the $\times$ notation: in general, $(\underline{a} \times \underline{b}) \times \underline{c} \neq \underline{a} \times (\underline{b} \times \underline{c})$.

The third form of multiplication is the *dyadic product*, or outer product, $\underline{Z} = \underline{a}\,\overline{\underline{b}}$, the result of which is a second-order tensor:

$$[\underline{Z}]^C = [\underline{a}]^C \overline{[\underline{b}]}^C = \begin{bmatrix} a_1 \\ a_2 \\ a_3 \end{bmatrix} \begin{bmatrix} b_1 & b_2 & b_3 \end{bmatrix} = \begin{bmatrix} a_1 b_1 & a_1 b_2 & a_1 b_3 \\ a_2 b_1 & a_2 b_2 & a_2 b_3 \\ a_3 b_1 & a_3 b_2 & a_3 b_3 \end{bmatrix}. \tag{2.29}$$

We will use this third form least commonly here, but it is conceptually useful as a way of constructing a basis of second-order tensors, and extending intuition from first-order tensors to second-order tensors.

Multiplication of Second-Order Tensors with First-Order Tensors

We can also compute the product $\underline{z} = \underline{A}\,\underline{b}$, where $\underline{A}$ is a second-order tensor, and $\underline{b}$ is a first-order tensor, and their product is a first-order tensor. Coordinating in some system C gives:

$$\underline{z} = \underline{A}\,\underline{b} \Leftrightarrow [\underline{z}]^C = [\underline{A}]^C [\underline{b}]^C. \tag{2.30}$$

Addition of Second-Order Tensors

The sum of two second-order tensors $\underline{A}$ and $\underline{B}$ is simply given by the sum of their components, in any allowable coordinate system C:

$$\underline{Z} := \underline{A} + \underline{B} \Leftrightarrow [\underline{Z}]^C := [\underline{A}]^C + [\underline{B}]^C \tag{2.31}$$

Product of Second-Order Tensors

The product of two second-order tensors $\underline{A}$ and $\underline{B}$ in any coordinate system C is simply given by their matrix product in C:

$$\underline{Z} := \underline{A}\,\underline{B} \Leftrightarrow [\underline{Z}]^C := [\underline{A}]^C [\underline{B}]^C. \tag{2.32}$$

[5] The name 'cross product' is again unfortunate, as it refers to an arbitrary multiplication symbol, while 'the vector product' is informative, indicating the operation's result.

2.3 Frames and Coordinate Systems

We think of a frame as an unambiguous collection of at least three points whose relative distances are fixed, and which do not lie in a line; and an associated rule to generate three orthogonal directions therefrom. For example, consider a (rigid) book, and define three points on the cover (three of the corners, say). Clearly, these are at a fixed distance from one another (assuming the book is rigid). Then define a series of directions, say $\underline{1}^A$, $\underline{2}^A$, and $\underline{3}^A$ as follows: let $\underline{1}^A$ point from the first corner, to the second. Let $\underline{2}^A$ be defined to lie in the plane of the page, orthogonal to $\underline{1}^A$, but pointing towards the third corner. The third direction, $\underline{3}^A$ follows from the vector product: $\underline{3}^A := \underline{K}(\underline{1}^A)\,\underline{2}^A$. We can now interchange the orientation of the frame (i.e., the book) with the description of the directions $\underline{1}^A$, $\underline{2}^A$, and $\underline{3}^A$. As we'd discussed earlier, we may interchangeably think of the coordinate system A, defined by the frame A. However, we may also define additional coordinate systems, also attached to the frame (e.g., one which is rotated around $\underline{3}^A$ by 45°).

When we model vehicles and their dynamics, we'll use reference points and reference frames. A reference point could be, for example, the center of mass, while the reference frame could be a body-fixed frame defined by three body-fixed vectors. Some important reference frames are:

body-fixed frame: $\underline{i}^B$ for a UAV, this is typically defined as a set of directions emanating from the vehicle's center of mass. We will *always* follow the robotics convention that $\underline{1}^B$ is forward, and $\underline{3}^B$ is *up*, rather than the aeronautical convention which has $\underline{3}^B$ pointing down.[6]

earth-fixed frame: $\underline{i}^E$ a local frame, attached to the ground, with $\underline{1}^E$ horizontal (e.g., north), and $\underline{3}^E$ pointing up to the sky. We will typically use this as an inertial frame, and it is sufficient for describing objects travelling over relatively short distances.

heliocentric frame: centered on the gravitational center of the sun (the sun is a ball of gas, what's a fixed point in the sun?), with one direction defined as pointing to a star that may be assumed stationary in the sky. The second direction may defined as perpendicular to the ecliptic (the plane along which the earth moves). This is used as an approximation for an inertial frame for applications such as interplanetary travel (e.g. science missions to Saturn).

Notice that we defined only two directions for each frame: because we're working with orthogonal, right-handed coordinate systems, we only need to specify two directions and the third follows unambiguously through the vector product.

[6] The intuition for the aerospace convention follows from the desire to have a positive pitch angle correspond to "nose up", but leads to a positive velocity along 3^B corresponding to a decreasing height.

2.3.1 Coordinate Transformation Matrix

We will often need to describe the same quantity in different coordinate systems. The most common example of this will be changing a description into the body-fixed coordinate system from the earth-fixed coordinate system (or vice-versa). Recalling the discussion in Sect. 2.1.2: this is a linear transformation, which we can express as a matrix multiplication, and we write the matrix as $[T]^{BA}$ to mean "the transformation matrix of coordinate system B with respect to A". Note that the transformation matrix is *not a tensor*, it is simply a matrix.

Revisiting the original derivation of the transformation matrix in (2.15), we recall that, for some tensor $\underline{x}$,

$$\left[\,\underline{x}\,\right]^B = [T]^{BA}\left[\,\underline{x}\,\right]^A \tag{2.33}$$

For the scalar coordinates $\left[\,\underline{x}\,\right]^A = \left(x_1^A, x_2^A, x_3^A\right)$ we had

$$\underline{x} = x_1^A\,\underline{1^A} + x_2^A\,\underline{2^A} + x_3^A\,\underline{3^A} \tag{2.34}$$

which we then coordinate in system B:

$$\left[\,\underline{x}\,\right]^B = x_1^A\left[\,\underline{1^A}\,\right]^B + x_2^A\left[\,\underline{2^A}\,\right]^B + x_3^A\left[\,\underline{3^A}\,\right]^B \tag{2.35a}$$

$$= \left[\left[\,\underline{1^A}\,\right]^B \left[\,\underline{2^A}\,\right]^B \left[\,\underline{3^A}\,\right]^B\right]\begin{bmatrix} x_1^A \\ x_2^A \\ x_3^A \end{bmatrix} \tag{2.35b}$$

Comparing (2.33) and (2.35b) we see that

$$[T]^{BA} = \left[\left[\,\underline{1^A}\,\right]^B \left[\,\underline{2^A}\,\right]^B \left[\,\underline{3^A}\,\right]^B\right] \tag{2.36}$$

and by a similar argument (but starting with $\left[\,\underline{x}\,\right]^B$) that

$$[T]^{BA} = \begin{bmatrix} \overline{\left[\,\underline{1^B}\,\right]^A} \\ \overline{\left[\,\underline{2^B}\,\right]^A} \\ \overline{\left[\,\underline{3^B}\,\right]^A} \end{bmatrix} \tag{2.37}$$

That is, the coordinate transformation matrix to the coordinate systems of B from that of A is formed by the expressing the base vectors of one frame in the coordinate system associated with the other.

Properties

Transformation matrices have many interesting properties, and much energy has been devoted to studying them. You can find a lot of information in Shuster (1993), but beware of differing notations, and incompatible definitions between sources. Below, m we list thtree important facts.

Fact 2.1 *Transformation matrices are orthogonal, that is,* $\overline{[T]}^{BA} = \left([T]^{BA}\right)^{-1} = [T]^{AB}$

Proof Let $\underline{x}$ be an arbitrary, nonzero, first-order tensor. Then

$$\left[\underline{x}\right]^B = [T]^{BA}\left[\underline{x}\right]^A \tag{2.38}$$

The length of the tensor is unchanged by the coordinate system used, so that we can write

$$\|\underline{x}\|^2 = \overline{\left[\underline{x}\right]}^A\left[\underline{x}\right]^A = \overline{\left[\underline{x}\right]}^B\left[\underline{x}\right]^B = \overline{\left[\underline{x}\right]}^A\overline{[T]}^{BA}[T]^{BA}\left[\underline{x}\right]^A \tag{2.39}$$

or

$$\overline{\left[\underline{x}\right]}^A\left(\left[\underline{I}\right]^A - \overline{[T]}^{BA}[T]^{BA}\right)\left[\underline{x}\right]^A = 0 \tag{2.40}$$

with $\left[\underline{I}\right]^A$ the matrix identity. Since (2.40) must hold for any $\underline{x}$, it follows that the term in brackets must be zero, i.e.

$$\left[\underline{I}\right] = \overline{[T]}^{BA}[T]^{BA} \quad \Leftrightarrow \quad \left([T]^{BA}\right)^{-1} = \overline{[T]}^{BA} \tag{2.41}$$

Finally, from (2.38), we get

$$\left[\underline{x}\right]^A = \left([T]^{BA}\right)^{-1}\left[\underline{x}\right]^B = [T]^{AB}\left[\underline{x}\right]^B \quad \Leftrightarrow \quad \left([T]^{BA}\right)^{-1} = \overline{[T]}^{BA} = [T]^{AB} \tag{2.42}$$

See also Problem 2.16. □

Fact 2.2 *Transformation matrices have determinant of* ± 1.

Proof Using Fact 2.1, we have

$$1 = \det\left(\left[\underline{I}\right]\right) = \det\left(\overline{[T]}^{AB}[T]^{AB}\right) \tag{2.43a}$$

$$= \det\left(\overline{[T]}^{AB}\right)\det\left([T]^{AB}\right) \tag{2.43b}$$

$$= \det\left([T]^{AB}\right)\det\left([T]^{AB}\right) \tag{2.43c}$$

$$= \det\left([T]^{AB}\right)^2 \tag{2.43d}$$

□

By convention, we consider those with determinant +1 to be "proper", since they describe right-handed coordinate systems.

Fact 2.3 *Transformations are composed as* $[T]^{AC} = [T]^{AB}\,[T]^{BC}$ *(note the contraction of superscripts).*

Proof Using the properties of matrix multiplication, for any $\underline{x}$ and coordinate systems A, B, and C we have

$$\begin{aligned}[\underline{x}]^A &= [T]^{AB}\,[\underline{x}]^B = [T]^{AB}\left([T]^{BC}\,[\underline{x}]^C\right) = \left([T]^{AB}\,[T]^{BC}\right)[\underline{x}]^C \\ &= [T]^{AC}\,[\underline{x}]^C\,. \end{aligned} \tag{2.44}$$

Since $\underline{x}$ was arbitrary, it follows that $[T]^{AB}\,[T]^{BC} = [T]^{AC}$. □

2.3.2 Euler angles

We often want to describe the change of coordinates in a compact way, more compact than the nine element matrix $[T]^{AB}$. A common way to describe this change is using Euler angles, of which there are many different conventions. Here we will only describe that which is most common in aerospace, the "3-2-1" yaw-pitch-roll angles, illustrated in Fig. 2.1.

To construct the transformation $[T]^{BE}$, we follow a series of simple steps. These steps will require the introduction of two intermediate frames, which we will write as X and Y, and we will construct from X from E, then Y from X, and finally B from Y: $[T]^{BE} = [T]^{BY}\,[T]^{YX}\,[T]^{XE}$. This is illustrated in Fig. 2.1.

In the first step, we generate the X coordinate system from by rotating the E coordinate system about the axis $\underline{3^E}$ by the yaw angle ψ. Note that this implies $\underline{3^X} = \underline{3^E}$.

We then rotate the X coordinate system about $\underline{2^X}$ by the pitch angle θ to generate the Y coordinate system, with now $\underline{2^Y} = \underline{2^X}$.

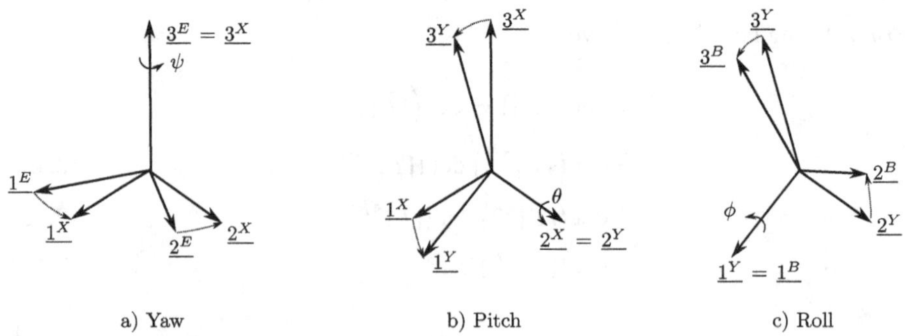

Fig. 2.1 The yaw-pitch-roll Euler sequence constructs the B coordinate system from the E coordinate system

Finally, we rotate Y by the roll angle ϕ about the $\underline{1}^Y$ axis to end up at the B coordinate system, with $\underline{1}^B = \underline{1}^Y$.

The component matrices can readily computed using (2.36):

$$[T]^{XE} = \begin{bmatrix} \cos\psi & \sin\psi & 0 \\ -\sin\psi & \cos\psi & 0 \\ 0 & 0 & 1 \end{bmatrix} \tag{2.45}$$

$$[T]^{YX} = \begin{bmatrix} \cos\theta & 0 & -\sin\theta \\ 0 & 1 & 0 \\ \sin\theta & 0 & \cos\theta \end{bmatrix} \tag{2.46}$$

$$[T]^{BY} = \begin{bmatrix} 1 & 0 & 0 \\ 0 & \cos\phi & \sin\phi \\ 0 & -\sin\phi & \cos\phi \end{bmatrix} \tag{2.47}$$

And tediously multiplying them yields:

$$[T]^{BE} = \begin{bmatrix} \mathrm{c}\psi\mathrm{c}\theta & \mathrm{s}\psi\mathrm{c}\theta & -\mathrm{s}\theta \\ \mathrm{c}\psi\mathrm{s}\theta\mathrm{s}\phi - \mathrm{s}\psi\mathrm{c}\phi & \mathrm{s}\psi\mathrm{s}\theta\mathrm{s}\phi + \mathrm{c}\psi\mathrm{c}\phi & \mathrm{c}\theta\mathrm{s}\phi \\ \mathrm{c}\psi\mathrm{s}\theta\mathrm{c}\phi + \mathrm{s}\psi\mathrm{s}\phi & \mathrm{s}\psi\mathrm{s}\theta\mathrm{c}\phi - \mathrm{c}\psi\mathrm{s}\phi & \mathrm{c}\theta\mathrm{c}\phi \end{bmatrix} \tag{2.48}$$

where we used the shorthand $\mathrm{c}x := \cos x$ and $\mathrm{s}x := \sin x$ so that the matrix fits on a page.

If, instead, we are given a transformation matrix with elements t_{ij}, we can compute the corresponding Euler angles as

$$\psi = \arctan\left(\frac{t_{12}}{t_{11}}\right), \qquad \theta = \arcsin\left(-t_{13}\right), \qquad \phi = \arctan\left(\frac{t_{23}}{t_{33}}\right) \tag{2.49}$$

With some experience, the Euler angles are very easy to interpret, and to visualize the orientation that corresponds to three particular angles (for example, by taking your right hand with thumb, index and middle finger extended, and sequentially rotating your hand). With the above expressions we can compute the corresponding rotation matrix, so that we can do useful computations. We can also do the inverse, for example when we have a simulation, and want to make a graph of a vehicle orientation: it is much easier to show the three angles!

Note—we will see other ways to encode rotation later, including the Euler Symmetric Parameters ("quaternion of rotation")—see Sect. 8.1.

2.4 Rotation Tensor

We've seen the displacement vector $\underline{s_{AB}}$, which describes the displacement of point A w.r.t. point B. The rotation tensor is a similar concept, which encodes the relative orientation of two frames, e.g. $\underline{R^{BA}}$ is the orientation of frame B w.r.t. frame A. This is different from the transformation matrix, in that the rotation tensor is a change in the physical world, while the transformation matrix only relates to our description of the world. As is a common theme, rotation will be somewhat similar to translation, but more complex!

Definition 2.3 If we have two frames A and B, with the three base directions $\underline{i^A}$ and $\underline{i^B}$ for $i \in \{1, 2, 3\}$, and some coordinate system C, the rotation tensor maps one to the other as

$$\left[\underline{i^B}\right]^C = \left[\underline{R^{BA}}\right]^C \left[\underline{i^A}\right]^C \quad \text{for} \quad i \in \{1, 2, 3\} \tag{2.50}$$

Fact 2.4 *The object* $\underline{R^{BA}}$ *of (2.50) is a tensor, i.e. for any coordinate systems C and D*

$$\left[\underline{R^{BA}}\right]^D = [T]^{DC} \left[\underline{R^{BA}}\right]^C \overline{[T]}^{DC} \tag{2.51}$$

Proof We begin by writing (2.50) in both coordinate systems C and D:

$$\left[\underline{i^B}\right]^C = \left[\underline{R^{BA}}\right]^C \left[\underline{i^A}\right]^C = \left[\underline{R^{BA}}\right]^C [T]^{CD} \left[\underline{i^A}\right]^D \tag{2.52}$$

$$\left[\underline{i^B}\right]^C = [T]^{CD} \left[\underline{i^B}\right]^D = [T]^{CD} \left[\underline{R^{BA}}\right]^D \left[\underline{i^A}\right]^D \tag{2.53}$$

from which follows

$$\left(\left[\underline{R^{BA}}\right]^C [T]^{CD} - [T]^{CD} \left[\underline{R^{BA}}\right]^D\right) \left[\underline{i^A}\right]^D = 0 \tag{2.54}$$

Since this has to hold for all three $\underline{i^A}$, we have

$$\left[\underline{R^{BA}}\right]^D = \left([T]^{CD}\right)^{-1} \left[\underline{R^{BA}}\right]^C [T]^{CD} = [T]^{DC} \left[\underline{R^{BA}}\right]^C \overline{[T]}^{DC}. \tag{2.55}$$

(*Note that this uses the insight that the three* $\left[\underline{i^A}\right]^D$ *span* $\mathbb{R}^3$.) □

Fact 2.5 *Rotations concatenate by multiplication:* $\underline{R^{CA}} = \underline{R^{CB}}\ \underline{R^{BA}}$.

The proof follows straightforwardly from the definiton.

Fact 2.6 *Rotations tensors are related to their coordinate systems through the transformation matrix:*

$$\left[\underline{R^{BA}}\right]^A = \left[\underline{R^{BA}}\right]^B = \overline{[T]}^{BA} = [T]^{AB} \tag{2.56}$$

Note that this is the same when expressed in either coordinate system B or A, and that it is the *transpose* of the transformation matrix.

Proof We note that for any frames A and B, with their coordinate systems, we have

$$[\underline{i}^B]^B = [\underline{i}^A]^A \quad i \in \{1, 2, 3\} \tag{2.57}$$

Using the definition of the rotation tensor ($\underline{i}^B = \underline{R}^{BA}\, \underline{i}^A$), we have

$$[\underline{i}^B]^A = [\underline{R}^{BA}]^A [\underline{i}^A]^A = [\underline{R}^{BA}]^A [\underline{i}^B]^B = [\underline{R}^{BA}]^A [T]^{BA} [\underline{i}^B]^A \tag{2.58}$$

or

$$\left([\underline{I}] - [\underline{R}^{BA}]^A [T]^{BA}\right) [\underline{i}^B]^A = [\underline{0}] \tag{2.59}$$

Since this has to hold over all three $[\underline{i}^B]^A$, which span $\mathbb{R}^3$, it follows that the product $[\underline{R}^{BA}]^A [T]^{BA}$ is identity, or $[\underline{R}^{BA}]^A = \overline{[T]}^{BA}$.

Coordinating instead the definition in system B, we would get $[\underline{R}^{BA}]^B = \overline{[T]}^{BA}$. □

From this fact, we can derive that the rotation tensor: (a) is orthogonal (i.e. $\underline{R}^{AB}\, \overline{\underline{R}^{AB}} = \underline{I}$), (b) has determinant $+1$, and (c) reverses the order of rotation if it is transposed, i.e. $\overline{\underline{R}^{BA}} = \underline{R}^{AB}$.

We can furthermore use the relationship to the transformation matrix to construct physical a rotation from a sequence of rotations (perhaps a sequence of easy-to-describe steps). We shall see, in Sect. 8.1.2, how Euler's theorem[7] states that we can describe *any* rotation as a rotation by an angle ϕ about a unit vector $\underline{n}$.[8]

Problems for Chapter 2

Problem 2.1 (*4 points*) Consider the triads for X and Y, as shown below. Compute the transformation matrix $[T]^{XY}$.

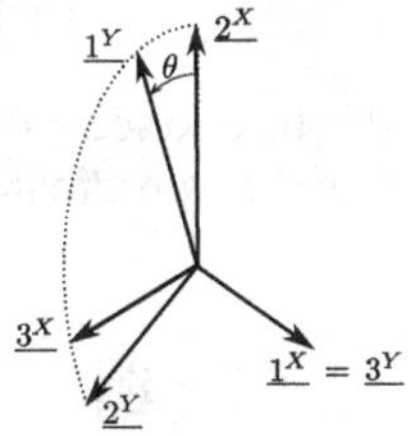

[7] Note: Euler had many theorems. This is only one of them.

[8] Sanity check: rotations have three degrees of freedom (e.g. yaw, pitch, roll). How many degrees of freedom in ϕ and $\underline{n}$? Does this make sense?.

Problem 2.2 (*6 points*) Given is the transformation matrix $[T]^{XY}$, below. Referring to the figure, relate $\underline{a}$, $\underline{b}$, and $\underline{c}$ to the tensors $\underline{1}^Y$, $\underline{2}^Y$, and $\underline{3}^Y$, with ξ as indicated in the figure. Motivate your answers.

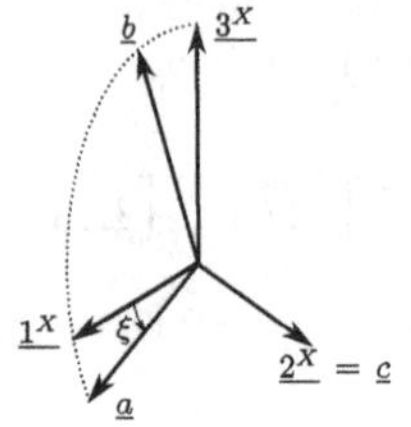

$$[T]^{XY} = \begin{bmatrix} 0 & \cos\xi & \sin\xi \\ 1 & 0 & 0 \\ 0 & -\sin\xi & \cos\xi \end{bmatrix}$$

Problem 2.3 (*3 points*) We define three points P, Q, and R with displacements as below, and the transformation between frames A and B.

$$\left[\underline{s_{PR}}\right]^A = \begin{bmatrix} -2 \\ 1 \\ 3 \end{bmatrix}, \quad \left[\underline{s_{QP}}\right]^A = \begin{bmatrix} 1 \\ 0 \\ -4 \end{bmatrix}, \quad [T]^{AB} = \begin{bmatrix} 0 & -1 & 0 \\ 1 & 0 & 0 \\ 0 & 0 & 1 \end{bmatrix}$$

Compute $\left[\underline{s_{QR}}\right]^B$.

Problem 2.4 (*4 points*) Consider the below matrix:

$$M = \begin{bmatrix} 0.8 & -0.8 & 0 \\ 0.8 & 0.8 & 0 \\ 0 & 0 & 1 \end{bmatrix}$$

Show that this matrix cannot be a transformation matrix.

Problem 2.5 (*2 points*) Numerically verify that $\overline{[T]}^{AB}\,[T]^{AB} = \left[\underline{I}\right]^A$, using

$$[T]^{AB} = \begin{bmatrix} 0 & 1 & 0 \\ -1 & 0 & 0 \\ 0 & 0 & 1 \end{bmatrix}.$$

Problem 2.6 (*4 points*) Let $\underline{X}$ and $\underline{Y}$ be two second order tensors. The tensor product of two second order tensors, in a given coordinate system C, is defined using the matrix product

$$[\underline{X}]^C\,[\underline{Y}]^C$$

Prove that this is a tensor operation, i.e. that there exists a second order tensor $\underline{Z}$ so that $\underline{Z} = \underline{XY}$.

Problem 2.7 (*4 points*) Let $\underline{x}$ and $\underline{y}$ be two first order tensors. The dyadic product of two first order tensors, in a given coordinate system C, is defined using the matrix product

$$\left[\underline{x}\right]^C \overline{\left[\underline{y}\right]}^C$$

Prove that this is a tensor operation, i.e. that there exists a second order tensor $\underline{Z}$ so that $\underline{Z} = \underline{x}\,\overline{\underline{y}}$.

Problem 2.8 (*4 points*) Show that, for any coordinate systems A and B, $\overline{\left[\underline{a}\right]}^A \left[\underline{b}\right]^A = \overline{\left[\underline{a}\right]}^B \left[\underline{b}\right]^B$ and that we can therefore write $\overline{\left[\underline{a}\right]}^A \left[\underline{b}\right]^A = \overline{\underline{a}}\,\underline{b}$.

Problem 2.9 (*4 points*) The sum of two first-order tensors (say, $\underline{x} + \underline{y}$ for some $\underline{x}$ and $\underline{y}$) is defined such that, in any coordinate system C, the result is the quantity $\left[\underline{z}\right]^C$ such that

$$\left[\underline{z}\right]^C := \left[\underline{x}\right]^C + \left[\underline{y}\right]^C$$

Prove that $\underline{z}$ is indeed a first-order tensor.

Problem 2.10 (*8 points*) Given two first-order tensors $\underline{x}$ and $\underline{y}$, show that $\underline{K}\left(\underline{x} + \underline{y}\right) = \underline{K}\left(\underline{x}\right) + \underline{K}\left(\underline{y}\right)$.

Problem 2.11 (*8 points*) Show that a change of coordinate systems does not change the length of a first order tensor. That is, $\left\|\left[\underline{x}\right]^A\right\| = \left\|\left[\underline{x}\right]^B\right\|$ for any coordinate systems A and B.

Hint: Recall that

$$\left\|\left[\underline{x}\right]^A\right\| = \sqrt{\overline{\left[\underline{x}\right]}^A \left[\underline{x}\right]^A} = \sqrt{(x_1^A)^2 + (x_2^A)^2 + (x_3^A)^2} \text{ for } \left[\underline{x}\right]^A = (x_1^A, x_2^A, x_3^A)$$

Use the orthogonality property of the transformation matrix.

(*Note that this is a circular argument, as we'd used the length to derive orthogonality. See, as alternative, Problem* 2.16.)

Problem 2.12 (*8 points*) This problem considers properties of the vector product of first-order tensors.

(a) Let $\underline{Y}$ be a second order tensor, and $\underline{x}$ a first-order tensor; and let A be some coordinate system. Let $\left[\underline{x}\right]^A$ be an eigenvector of $\left[\underline{Y}\right]^A$, with associated eigenvalue λ, so that $\left[\underline{Y}\right]^A \left[\underline{x}\right]^A = \lambda \left[\underline{x}\right]^A$. Show that $\underline{x}$ is an "eigen-first-order-tensor" of $\underline{Y}$, i.e. show show that $\underline{Y}\underline{x} = \lambda\underline{x}$ (recall, this means the relationship has to hold when coordinated in *any* coordinate system).

(b) Given is the first-order tensor $\underline{x}$ and the coordinate systems A, so that $\left[\underline{x}\right]^A = (x_1, x_2, x_3)$. Let $\underline{X} = \underline{K}(\underline{x})$. Show that $\left[\underline{x}\right]^A$ is an eigenvector of $\left[\underline{X}\right]^A$, with corresponding eigenvalue 0.

From the above two points, we can conclude that $\underline{K}(\underline{x})\,\underline{x} = \underline{0}$. How does this relate to your previous knowledge of the vector (cross) product?

Problem 2.13 (*5 points*) Let F be some frame with base directions $\underline{1^F}$, $\underline{2^F}$, and $\underline{3^F}$. Consider the three first-order tensors $\underline{x} = x_1\,\underline{1^F} + x_2\,\underline{2^F} + x_3\,\underline{3^F}$, $\underline{y} = y_1\,\underline{1^F} + y_2\,\underline{2^F} + y_3\,\underline{3^F}$, and $\underline{z} = z_1\,\underline{1^F} + z_2\,\underline{2^F} + z_3\,\underline{3^F}$. In a previous course, say Mechanics II, you may have learned the vector ("cross") product of $\underline{x}$ and $\underline{y}$ as

$$\underline{x} \times \underline{y} := \det\left(\begin{bmatrix} \underline{1^F} & \underline{2^F} & \underline{3^F} \\ x_1 & x_2 & x_3 \\ y_1 & y_2 & y_3 \end{bmatrix}\right) \tag{2.60}$$

where det is the matrix determinant.

(a) Write the coordinated vectors $\left[\underline{y}\right]^F$ and $\left[\underline{z}\right]^F$.

(b) Using the determinant formula above, compute $\left[\underline{x} \times \underline{y}\right]^F$

(c) Using the definitions from this book with $\underline{X} := \underline{K}(\underline{x})$ and $\underline{Y} := \underline{K}(\underline{y})$, compute $\left[\underline{X}\right]^F$ and the result $\left[\underline{X}\right]^F\left[\underline{y}\right]^F$.

(d) For arbitrary $\underline{w}$, show that the cross product, defined using the determinant, is not associative, that is show that $\underline{x} \times \left(\underline{y} \times \underline{w}\right) \neq \left(\underline{x} \times \underline{y}\right) \times \underline{w}$.

(e) Show that the vector product we defined *is* associative, that is show that

$$\left[\underline{X}\right]^F\left(\left[\underline{Y}\right]^F\left[\underline{w}\right]^F\right) = \left(\left[\underline{X}\right]^F\left[\underline{Y}\right]^F\right)\left[\underline{w}\right]^F$$

Note that, therefore we can unambiguously write it simply as the product $\underline{X}\,\underline{Y}\,\underline{w}$.

Problem 2.14 (*8 points*) We wish to investigate whether a proposed method of multiplying two first-order tensors generates a second-order tensor. We will denote the proposed product with the symbol $\otimes$. For the definition, we write the first-order tensors $\underline{x}$ and $\underline{y}$ in some coordinate system C, so that $\left[\underline{x}\right]^C = (x_1^C, x_2^C, x_3^C)$ and $\left[\underline{y}\right]^C = (y_1^C, y_2^C, y_3^C)$. The proposed product then follows as

$$\left[\underline{x}\right]^C \otimes \left[\underline{y}\right]^C = \begin{bmatrix} x_1^C y_1^C & 0 & 0 \\ 0 & x_2^C y_2^C & 0 \\ 0 & 0 & x_3^C y_3^C \end{bmatrix}$$

Is this a valid tensor operation? In other words, does there exist some second order tensor $\underline{Z}$ so that $\underline{Z} = \underline{x} \otimes \underline{y}$?

If the statement is true, provide a proof; if false, provide a counter-example.

Problem 2.15 (*8 points*) Consider the following operation, which maps a $\mathbb{R}^3$ vector into a $\mathbb{R}^{3\times3}$ matrix. Let $\left[\, \underline{x} \,\right] = (x_1, x_2, x_3)$.

$$\operatorname{diag}(\left[\, \underline{x} \,\right]) := \begin{bmatrix} x_1 & 0 & 0 \\ 0 & x_2 & 0 \\ 0 & 0 & x_3 \end{bmatrix}$$

Does this define a second-order tensor? I.e., given some first-order tensor $\underline{v}$, does $\operatorname{diag}(\underline{v})$ define a second-order tensor?

If the statement is true, provide a proof; if false, provide a counter-example.

Problem 2.16 (*8 points*) Consider two points in the world, P and Q, with displacement $\underline{s_{PQ}}$ and a frame C, so that $\left[\, \underline{s_{PQ}} \,\right]^C = (x, y, z)$. Starting only with Pythagoras' theorem show that the length of distance between point P and Q, that is, $\left\| \underline{s_{PQ}} \right\|$, can be measured with a ruler, or computed from coordinates as the Euclidean norm $\left\| \left[\, \underline{s_{PQ}} \,\right]^C \right\|$

$$\left\| \underline{s_{PQ}} \right\| = \left\| \left[\, \underline{s_{PQ}} \,\right]^C \right\|$$

Recall Pythagoras' theorem: *the sum of the squares of the lengths of the orthogonal sides of a right triangle equal the square of the length of the remaining side*; and the definition of the Euclidean norm of a vector:

$$\left\| \begin{bmatrix} a \\ b \\ c \end{bmatrix} \right\| := \sqrt{a^2 + b^2 + c^2}$$

(*It may be helpful to start with a simple sketch showing* $\underline{s_{PQ}}$, *the frame* C, *and the coordinates* x, y, z.)

Problem 2.17 (*4 points*) Consider first-order tensor $\underline{x}$, and two coordinate systems C and D. We claimed that we can generate a related second-order, skew-symmetric tensor: $\underline{X} = \underline{K}(\underline{x})$. Using

$$[T]^{DC} = \begin{bmatrix} 1 & 0 & 0 \\ 0 & \cos\phi & \sin\phi \\ 0 & -\sin\phi & \cos\phi \end{bmatrix} \tag{2.61}$$

and $\left[\, \underline{x} \,\right]^C = (x_1, x_2, x_3)$ verify that this is a valid second-order tensor; that is, show that

$$\left[\, \underline{K}(\underline{x}) \right]^C = [T]^{CD} \left[\, \underline{K}(\underline{x}) \right]^D \overline{[T]}^{CD} \tag{2.62}$$

with

$$[\underline{K}(\underline{x})]^D = K\left([\underline{x}]^D\right) = K\left([T]^{DC}[\underline{x}]^C\right) \tag{2.63}$$

Problem 2.18 *(10 points)*

In this problem, we prove that the skew-symmetric operator $\underline{K}(\underline{\cdot})$ defined in Sect. 2.2.2, specifically (2.20), defines a second-order tensor.

To do this, we first define an alternative method for computing a vector product—an operation completed using ruler and protractor, without any linear algebra. Let $\underline{x}$ and $\underline{y}$ be two first-order tensors, and define the first-order tensor $\underline{z}$ as their *vector product*

$$\underline{z} := \underline{\nu}\left(\underline{x}, \underline{y}\right) \tag{2.64}$$

with the function $\underline{\nu}(\cdot,\cdot)$ defined as generating a first-order tensor $\underline{z}$ having the following properties:

- $\underline{z}$ is orthogonal to both $\underline{x}$ and $\underline{y}$.
- The direction of $\underline{z}$ is defined using the right-hand rule (aligning the thumb of the right-hand with $\underline{z}$, and the index finger with $\underline{x}$ allows one to sweep the index finger to $\underline{y}$).
- The magnitude of $\underline{z}$ is defined as $\|\underline{z}\| = \|\underline{x}\| \, \|\underline{y}\| \sin(\alpha)$ where $\alpha = \angle\left(\underline{x}, \underline{y}\right)$ is the angle between $\underline{x}$ and $\underline{y}$.

Clearly, $\underline{z}$ is a first-order tensor (it is defined as a physical object in space, without regards to any specific coordinate system). Next, let C be an arbitrary coordinate system, so that $[\underline{x}]^C = (x_1, x_2, x_3)$ and $\left[\underline{y}\right]^C = (y_1, y_2, y_3)$.

a. Show that the vector $[\underline{v}]$

$$[\underline{v}] = \begin{bmatrix} 0 & -x_3 & x_2 \\ x_3 & 0 & -x_1 \\ -x_2 & x_1 & 0 \end{bmatrix} \begin{bmatrix} y_1 \\ y_2 \\ y_3 \end{bmatrix}$$

satisfies all three properties of $\underline{z}$; i.e., $[\underline{v}] = [\underline{z}]^C$.

We can thus write $[\underline{z}]^C = K\left([\underline{x}]^C\right)\left[\underline{y}\right]^C$ where we defined the (purely linear-algebra) operator $K(\cdot)$ which maps from $\mathbb{R}^3$ to $\mathbb{R}^{3\times 3}$ as

$$K\left(\begin{bmatrix} a \\ b \\ c \end{bmatrix}\right) = \begin{bmatrix} 0 & -c & b \\ c & 0 & -a \\ -b & a & 0 \end{bmatrix} \tag{2.65}$$

Noting that C was arbitrary, we can conclude that, for some other coordinate system D, we also have $[\underline{z}]^D = K\left([\underline{x}]^D\right)\left[\underline{y}\right]^D$.

b. Prove now that $K\left(\left[\underline{x}\right]^{D}\right) = K\left([T]^{DC}\left[\underline{x}\right]^{C}\right) = [T]^{DC}\, K\left(\left[\underline{x}\right]^{C}\right)\overline{[T]}^{DC}$.

From this we conclude that there is a second-order tensor object $\underline{\underline{K}}(\underline{x})$ defined using the linear algebra operator $K(\cdot)$ as

$$\left[\underline{\underline{K}}(\underline{x})\right]^{C} = K\left(\left[\underline{x}\right]^{C}\right) = \begin{bmatrix} 0 & -x_3 & x_2 \\ x_3 & 0 & -x_1 \\ -x_2 & x_1 & 0 \end{bmatrix} \text{ for } \left[\underline{x}\right]^{C} = \begin{bmatrix} x_1 \\ x_2 \\ x_3 \end{bmatrix}$$

Problem 2.19 (*4 points*) Let $\underline{\underline{A}}$ be a second-order tensor, that is skew-symmetric, i.e., $\overline{\underline{\underline{A}}} = -\underline{\underline{A}}$. Prove that there exists a unique first-order tensor $\underline{b}$ so that $\underline{\underline{A}} = \underline{\underline{K}}\left(\underline{b}\right)$.

Note that you have to first prove that such a $\underline{b}$ *exists at all, and then that it is unique. Hint: a good strategy when asked to prove something like this, which seems painfully obvious, is a* proof by contradiction. *To do this, assume that the premise is false, and then show that a contradiction follows.*

Problem 2.20 (*10 points*) In this problem, we prove the linearity property of the scalar product (2.2) and (2.3).

Let $\underline{x}$, $\underline{y}$, and $\underline{z}$ be first-order tensors; and let α be a scalar. Starting with the definition (2.1), show:

a. $\sigma\left(\underline{x}, \underline{y}\right) = \sigma\left(\underline{y}, \underline{x}\right)$

b. $\sigma\left(\underline{x}, \alpha\underline{y}\right) = \alpha\sigma\left(\underline{x}, \underline{y}\right)$

For the last part of this problem, we will simplify to the 2D plane (generalizing to 3D requires diagrams that are annoying to sketch). Sketch out a diagram with distinct vectors $\underline{x}$, $\underline{y}$, and $\underline{z}$.

Define the three scalar lengths, and mark them on your diagram:

$$\begin{aligned} l_1 &= \left\|\underline{y}\right\| \cos\left(\angle\left(\underline{x}, \underline{y}\right)\right) \\ l_2 &= \left\|\underline{z}\right\| \cos\left(\angle\left(\underline{x}, \underline{z}\right)\right) \\ l_3 &= \left\|\underline{y} + \underline{z}\right\| \cos\left(\angle\left(\underline{x}, \underline{y} + \underline{z}\right)\right) \end{aligned}$$

c. Referring to your diagram, show that $l_1 + l_2 = l_3$.

By pre-multiplying this with $\left\|\underline{x}\right\|$, we conclude $\sigma\left(\underline{x}, \underline{y} + \underline{z}\right) = \sigma\left(\underline{x}, \underline{y}\right) + \sigma\left(\underline{x}, \underline{z}\right)$.

Problem 2.21 (*6 points*) Consider a plane (i.e., a 2D surface) with the normal unit-tensor $\underline{n}$ (i.e., $\left\|\underline{n}\right\| = 1$). Let $\underline{v}$ be some first-order tensor. We are interested in describing the *projection of the first-order tensor onto the plane*. For intuition, think of it as the shadow the tensor casts onto a surface perpendicular to a light shining in the direction $\underline{n}$.

For example, in a coordinate system C which has $\underline{3}^C = \underline{n}$, and where $\left[\underline{v}\right]^C = (v_1, v_2, v_3)$, then the projection of $\underline{v}$ onto the plane (call it $\underline{w}$) is

$$\left[\underline{w}\right]^C = \begin{bmatrix} v_1 \\ v_2 \\ 0 \end{bmatrix}$$

Show that this can be expressed as a tensor operation, and derive the second-order tensor $\underline{A}$ as a function only of the normal $\underline{n}$, so that $\underline{w} = \underline{A}\,\underline{v}$.

Problem 2.22 (*6 points*) Consider a mirror, defined as a plane with the normal unit-tensor $\underline{n}$ (i.e., $\|\underline{n}\| = 1$). Let $\underline{v}$ be some first-order tensor. We are interested in the *reflection of* $\underline{v}$ *in the mirror*.

For example, in a coordinate system C which has $\underline{3}^C = \underline{n}$, and where $\left[\underline{v}\right]^C = (v_1, v_2, v_3)$, then the reflection of $\underline{v}$ in the mirror (call it $\underline{w}$) is

$$\left[\underline{w}\right]^C = \begin{bmatrix} v_1 \\ v_2 \\ -v_3 \end{bmatrix}$$

Show that this can be expressed as a tensor operation, and derive the second-order tensor $\underline{A}$ as a function only of the normal $\underline{n}$, so that $\underline{w} = \underline{A}\,\underline{v}$.

Problem 2.23 (*6 points*) A UAV uses its on-board sensors to measure the displacement of a known landmark with respect to itself, expressed in its body-fixed coordinate system B, as $\left[\underline{s_{LB}}\right]^B$. From existing maps, it knows the displacement of the landmark with respect to the UAVs landing site H, expressed in the earth-fixed coordinate system E as $\left[\underline{s_{LH}}\right]^E$. The UAV also has an estimate of its orientation with respect to the earth, $[T]^{BE}$.

Provide an expression for the displacement of the UAV with respect to the landing site, expressed in the earth-fixed coordinate system. Your final result must use *only* known quantities.

Problem 2.24 (*10 points*) Consider a quadcopter carrying a load, as shown in the figure, with three points: B (a quadcopter's center of mass), L (the load's center of the mass) and T (the target on the ground), with relative displacements given by, e.g., $\underline{s_{BL}}$.

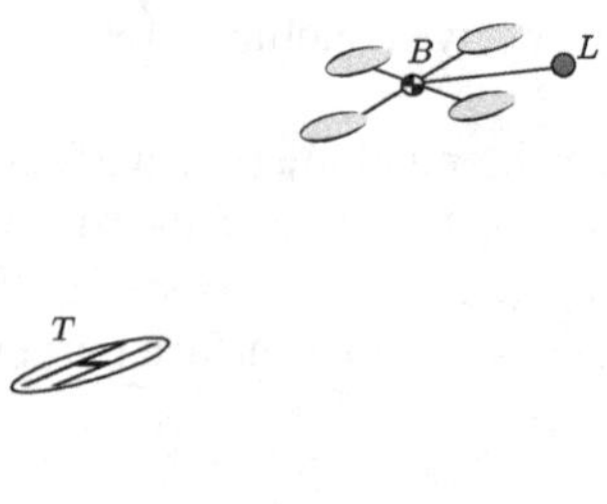

Given are two coordinate systems, E (earth-fixed) and B (body-fixed); and specifically:

$$[T]^{BE} = \begin{bmatrix} \frac{\sqrt{2}}{2} & \frac{\sqrt{2}}{2} & 0 \\ -\frac{\sqrt{2}}{2} & \frac{\sqrt{2}}{2} & 0 \\ 0 & 0 & 1 \end{bmatrix}$$

A ground-based radar knows the displacement of the quadcopter from the target; and the quadcopter knows the displacement of the load from its center of mass, specifically:

$$\left[\underline{s}_{CT}\right]^E = \begin{bmatrix} 25 \\ 12 \\ 50 \end{bmatrix} \text{m; and } \left[\underline{s}_{LC}\right]^B = \begin{bmatrix} 1 \\ 0 \\ 0 \end{bmatrix} \text{m.}$$

Compute the distance from the load to the target.

Problem 2.25 (*5 points*) The inertial measurement unit (IMU) on a quadcopter is mounted so that it returns measurements in the S coordinate system, shown in the figure below. The quadcopter's controller, however, is programmed in the B coordinate system, also shown.

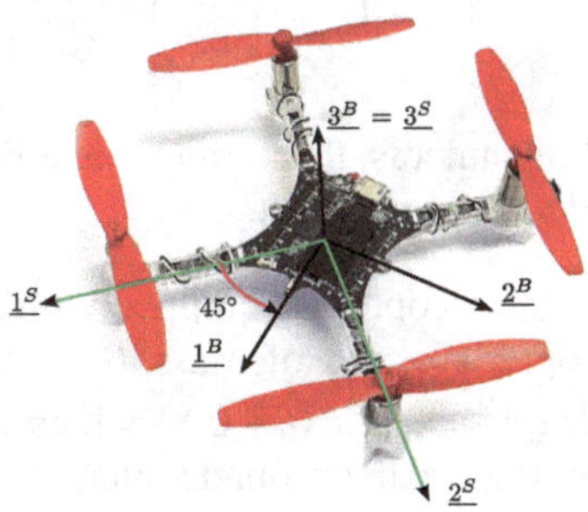

The IMU outputs the measurement $\left[\underline{\alpha}\right]^S = (5, 0, 10)\,\text{m/s}^2$. Convert this to a measurement that the controller can directly use, i.e. $\left[\underline{\alpha}\right]^B$.

Problem 2.26 (*10 points*) The orientation of a quadcopter B w.r.t. the earth E is described by the Euler angles $\psi = 90°$, $\theta = 45°$, $\phi = 30°$. It is flying through a uniform magnetic field, where the magnetic field vector is given by $\underline{\mu}$, and specifically:

$$\left[\underline{\mu}\right]^E = \begin{bmatrix} 1 \\ 0 \\ 0 \end{bmatrix}$$

The quadcopter is equipped with a field sensor, that measures $\left[\underline{\mu}\right]^B$. Compute the output of the sensor, $\left[\underline{\mu}\right]^B$, i.e. the magnetic field expressed in the body-fixed coordinate system. Use the transformation matrices defined for the Euler angles in (2.48).

Problem 2.27 (*10 points*) A quadcopter is equipped with a camera, which is located at the point of the quadcopter's center of mass B. The camera can rotate w.r.t. the quadcopter frame through two angles, equivalent to the Euler yaw ψ and pitch θ rotations (but no roll). The angles are constrained so that $\theta \in [-90°, 90°]$ and $\psi \in [0, 360°]$. We introduce the B frame, fixed w.r.t. the quadcopter, and the C frame fixed to the camera, so that

$$[T]^{CB} = \begin{bmatrix} \cos\theta & 0 & -\sin\theta \\ 0 & 1 & 0 \\ \sin\theta & 0 & \cos\theta \end{bmatrix} \begin{bmatrix} \cos\psi & \sin\psi & 0 \\ -\sin\psi & \cos\psi & 0 \\ 0 & 0 & 1 \end{bmatrix}$$

The camera points in the direction of $\underline{1}^C$.

Compute the needed angles ψ and θ so that the camera points at a target T located at

$$\left[\underline{s_{TC}}\right]^B = \begin{bmatrix} 0 \\ 5 \\ -5 \end{bmatrix} \text{m}$$

As an aside, in the context of cameras, this arrangement is called "pan-tilt", with the angle ψ being the pan, and θ the camera tilt.

Problem 2.28 (*8 points*) A quadcopter is equipped with a camera, which is located at the point of the quadcopter's center of mass B, and whose direction of view points along the direction $\left[\underline{c}\right]^B = (0.6, 0, -0.8)$ which is fixed with respect to the quadcopter. The frame B is attached to the quadcopter, and E is an Earth-fixed frame. The goal is to film a point P fixed on the ground, by moving the quadcopter, and specifically:

- have the point P at the center of the camera's field of view
- hover at a height of 15 m, with zero yaw.

Compute the required position of the quadcopter, relative to the point fixed on the ground, expressed in the Earth-fixed coordinate system E: $\left[\underline{s_{BP}}\right]^E$.

References

Lee T, Leok M, McClamroch NH (2017) Global formulations of Lagrangian and Hamiltonian dynamics on manifolds. Springer 13:31

Shuster MD (1993) A survey of attitude representations. J Astron Sci 41(4):439—517

Zipfel PH (2007) Modeling and simulation of aerospace vehicle dynamics, 2nd edn. AIAA (2007)

References

Chapter 3
Kinematics

★★★★★ *A very moving chapter.*

—Anonymous reviewer

Kinematics deals with how quantities change over time, e.g. the relationship between position and velocity. In the chapter after this we consider dynamics, i.e. how things are affected by forces and torques. We will notice that we have to be very pedantic about time derivatives. This chapter again follows Zipfel (2007) very closely.

3.1 Rotational Time Derivative

We want to find a time derivative operator that is coordinate free, and that allows us to keep working with tensors. To see why this is non-trivial, consider expressing a vector $\underline{s}$ in two coordinate systems A and B:

$$\left[\underline{s}\right]^A = [T]^{AB}\left[\underline{s}\right]^B \tag{3.1}$$

We define the element-wise time derivative of a coordinated vector $\left[\underline{x}\right]^A = (x_1, x_2, x_3)$ as

$$\left[\underline{\frac{\mathrm{d}}{\mathrm{d}t}x}\right]^A = \frac{\mathrm{d}}{\mathrm{d}t}\left[\underline{x}\right]^A = \begin{bmatrix} \frac{\mathrm{d}}{\mathrm{d}t}x_1 \\ \frac{\mathrm{d}}{\mathrm{d}t}x_2 \\ \frac{\mathrm{d}}{\mathrm{d}t}x_3 \end{bmatrix} \tag{3.2}$$

M. W. Mueller, *Dynamics and Control of Autonomous Flight*, Mathematical Engineering,
https://doi.org/10.1007/978-3-031-91871-1_3

i.e. we read $\left[\frac{\mathrm{d}}{\mathrm{d}t}s\right]^A$ as the time derivative applied to each scalar element of $\left[\underline{s}\right]^A$. If we take the time derivative of both sides of (3.1), and apply the product rule, we get:

$$\left[\frac{\mathrm{d}}{\mathrm{d}t}s\right]^A = [T]^{AB}\left[\frac{\mathrm{d}}{\mathrm{d}t}s\right]^B + \left[\frac{\mathrm{d}}{\mathrm{d}t}T\right]^{AB}\left[\underline{s}\right]^B \tag{3.3a}$$

$$\neq [T]^{AB}\left[\frac{\mathrm{d}}{\mathrm{d}t}s\right]^B \tag{3.3b}$$

In other words, if we do a time derivative of a list of coordinates, the result is not a tensor.

Instead, we will use the *rotational time derivative* as defined by Zipfel (2007), which requires specifying a frame. That is, we must specify from whose perspective we are taking the time derivative.

Definition 3.1 The rotational time derivative of a first-order tensor $\underline{x}$ w.r.t. a frame A and expressed in a coordinate system B is defined by

$$\left[\underline{D^A x}\right]^B := \left[\frac{\mathrm{d}}{\mathrm{d}t}x\right]^B + [T]^{BA}\overline{\left[\frac{\mathrm{d}}{\mathrm{d}t}T\right]^{BA}}\left[\underline{x}\right]^B \tag{3.4}$$

Notice that implicit in this is the coordinate system A associated with frame A.

If we set frame B to be the same as A, we get

$$\left[\underline{D^A x}\right]^A = \left[\frac{\mathrm{d}}{\mathrm{d}t}x\right]^A + [T]^{AA}\overline{\left[\frac{\mathrm{d}}{\mathrm{d}t}T\right]^{AA}}\left[\underline{x}\right]^A = \left[\frac{\mathrm{d}}{\mathrm{d}t}x\right]^A \tag{3.5}$$

because $[T]^{AA}$ is the identity, and its time derivative is zero. In other words, if we coordinate in the same frame, the rotational time derivative is the same as the ordinary time derivative.

To show that this is an invariant operator (i.e., generates a tensor), we refactor (3.3a):

$$\left[\underline{D^A s}\right]^A = \left[\frac{\mathrm{d}}{\mathrm{d}t}s\right]^A = [T]^{AB}\left[\frac{\mathrm{d}}{\mathrm{d}t}s\right]^B + \left[\frac{\mathrm{d}}{\mathrm{d}t}T\right]^{AB}\left[\underline{s}\right]^B \tag{3.6a}$$

$$= [T]^{AB}\left(\left[\frac{\mathrm{d}}{\mathrm{d}t}s\right]^B + [T]^{BA}\left[\frac{\mathrm{d}}{\mathrm{d}t}T\right]^{AB}\left[\underline{s}\right]^B\right) \tag{3.6b}$$

$$= [T]^{AB}\left[\underline{D^A s}\right]^B \tag{3.6c}$$

which means that $\underline{D^A s}$ transforms like a first-order tensor, and is therefore invariant.

It is useful to think of the specified frame as the *perspective from which the change is perceived*, in other words, we may read $D^A x$ as "how an observer attached to frame A perceives the change in the tensor $\underline{x}$".[1]

Definition 3.2 The rotational time derivative of a second-order tensor $\underline{X}$ w.r.t. a frame A and expressed in a coordinate system B is defined by

$$[\underline{D^A X}]^B := \left[\underline{\frac{\mathrm{d}}{\mathrm{d}t} X}\right]^B + [T]^{BA} \overline{\left[\frac{\mathrm{d}}{\mathrm{d}t} T\right]^{BA}} [\underline{X}]^B + [\underline{X}]^B \overline{[T]^{BA}} \left[\frac{\mathrm{d}}{\mathrm{d}t} T\right]^{BA} \tag{3.7}$$

It can also be shown that $\underline{D^A X}$ is a second-order tensor, but we will not spend more time on it.

The rotational time derivative has the following properties:

Linearity for constant scalars α and β, and tensors $\underline{x}$ and $\underline{y}$

$$D^A \left(\alpha \underline{x} + \beta \underline{y}\right) = \alpha D^A \underline{x} + \beta D^A \underline{y} \tag{3.8}$$

Product rule for tensors $\underline{X}$ and $\underline{y}$, and scalar w:

$$D^A \left(w \underline{X} \underline{y}\right) = \left(\frac{\mathrm{d}}{\mathrm{d}t} w\right) \underline{X} \underline{y} + w \left(D^A \underline{X}\right) \underline{y} + w \underline{X} \left(D^A \underline{y}\right) \tag{3.9}$$

The product rule also holds for products of two first-order tensors, etc.

3.2 Translational Velocity and Acceleration

Now that we are in a position to talk about time derivatives, we can introduce velocity and acceleration. Let $\underline{s_{BA_1}}$ be the displacement of point B w.r.t. point A_1, where A_1 is a fixed point on the frame A.

Definition 3.3 We define the velocity of point B w.r.t. the *frame* A as

$$\underline{v_B^A} := D^A \underline{s_{BA_1}} \tag{3.10}$$

Note that it doesn't matter which point A_1 on frame A we choose (see Problem 3.1). Clearly, by the properties of the rotational time derivative, $\underline{v_B^A}$ is a first-order tensor.

[1] The whole controversy about the relative motion of the sun and the earth could have been avoided had Galileo and the pope simply been clear about the frame with respect to which they were taking derivatives.

Definition 3.4 Similarly, we can define the translational acceleration of point B w.r.t. frame A as

$$\underline{a_B^A} := D^A D^A \underline{s_{BA_1}} = D^A \underline{v_B^A} \tag{3.11}$$

Again, this is a first-order tensor, by the properties of the rotational time derivative.

3.3 Angular Velocity

Whereas the translational velocity relates to the rate of change of displacement of two points, the angular velocity relates to the rate of change of orientation of two frames.

Definition 3.5 The angular velocity tensor of frame B w.r.t. frame A is defined as

$$\underline{\Omega^{BA}} := \left(D^A \underline{R^{BA}}\right) \overline{\underline{R^{BA}}} \tag{3.12}$$

This is a skew-symmetric tensor, that is, $\underline{\Omega^{BA}} + \overline{\underline{\Omega^{BA}}} = 0$. We can show this by starting with the observation

$$\underline{I} = \underline{R^{BA}}\,\overline{\underline{R^{BA}}} \tag{3.13}$$

Taking the time derivative with respect to frame A of both sides yields:

$$\begin{aligned}
\underline{0} = D^A\left(\underline{I}\right) &= D^A\left(\underline{R^{BA}}\,\overline{\underline{R^{BA}}}\right) && (3.14a)\\
&= \left(D^A \underline{R^{BA}}\right)\overline{\underline{R^{BA}}} + \underline{R^{BA}}\left(D^A \overline{\underline{R^{BA}}}\right) && (3.14b)\\
&= \left(D^A \underline{R^{BA}}\right)\overline{\underline{R^{BA}}} + \overline{\left(D^A \underline{R^{BA}}\right)\overline{\underline{R^{BA}}}} && (3.14c)\\
&= \underline{\Omega^{BA}} + \overline{\underline{\Omega^{BA}}} && (3.14d)
\end{aligned}$$

Since it is skew symmetric, it has only three degrees of freedom, and in some given coordinate system C we can express it using three scalars in a matrix:

$$\left[\underline{\Omega^{BA}}\right]^C = \begin{bmatrix} 0 & -\omega_3 & \omega_2 \\ \omega_3 & 0 & -\omega_1 \\ -\omega_2 & \omega_1 & 0 \end{bmatrix} \tag{3.15}$$

We can thus associate a unique first-order tensor to this, $\underline{\omega}^{BA}$, which is expressed in C as $\left[\underline{\omega}^{BA}\right]^C = (\omega_1, \omega_2, \omega_3)$. See Problem 2.19 to prove why this is unique. In other words, we have

$$\underline{\Omega}^{BA} = \underline{K}\left(\underline{\omega}^{BA}\right). \tag{3.16}$$

3.3.1 Orientation Differential Equations

Very often, it is of interest to describe how the orientation of two bodies change relative to one another, as a function of the angular velocities. For example, we may know the angular velocity of the body B with respect to the earth E from a rate gyroscope, which measures quantities in the body-fixed coordinate system: $\left[\underline{\omega}^{BE}\right]^B$. This is most often written as $\left[\underline{\omega}^{BE}\right]^B = (p, q, r)$, with the components called respectively the *roll rate*, *pitch rate*, and *yaw rate*.[2]

The challenge is now to use this to describe the orientation $\underline{R}^{BA}$, and the associated transformation matrix $[T]^{BA}$. This is needed, for example, if you are creating a simulation of a vehicle, or estimating an orientation based on data.

Starting from the definition of the angular velocity (3.12), we have

$$D^B\, \underline{R}^{EB} = \underline{\Omega}^{EB}\, \underline{R}^{EB} \tag{3.17}$$

Coordinating in B gives

$$\left[\frac{\mathrm{d}}{\mathrm{d}t}R^{EB}\right]^B = \left[\underline{D^B R^{EB}}\right]^B = \left[\underline{\Omega}^{EB}\right]^B \left[\underline{R}^{EB}\right]^B \tag{3.18a}$$

$$= \overline{\left[\underline{\Omega}^{BE}\right]}^B \left[\underline{R}^{EB}\right]^B \tag{3.18b}$$

Taking the transpose of both sides gives:

$$\left[\frac{\mathrm{d}}{\mathrm{d}t}R^{BE}\right]^B = \left[\underline{R}^{BE}\right]^B \left[\underline{\Omega}^{BE}\right]^B \tag{3.19}$$

Using the fact that $\left[\underline{R}^{BA}\right]^A = \left[\underline{R}^{BA}\right]^B = \overline{[T]}^{BA} = [T]^{AB}$, we get

$$\left[\frac{\mathrm{d}}{\mathrm{d}t}T\right]^{EB} = [T]^{EB}\left[\underline{\Omega}^{BE}\right]^B \tag{3.20}$$

[2] Note that, confusingly, these are not (in general) equal to the time derivative of the Euler roll-pitch-yaw angles, see (3.23). Getting these confused is a typical mistake of the inexperienced GNC engineer.

We note that this is a set of *nine* coupled differential equations, relating the scalar components of the transformation matrix to the angular velocity components.

Solution for Constant Angular Velocity

Recall that, for a scalar quantity x, which solves the ODE $\dot{x} = ax$, the solution is $x(t) = \exp(a)\, x(0)$.

If $\underline{\omega}^{BE}$ is constant then (3.20) is an autonomous linear differential equation, which we can solve for in closed form. Let $\left[\underline{T(t)}\right]^{EB}$ be the value of $[T]^{EB}$ at time t, then

$$\left[\underline{T(t_0 + \Delta t)}\right]^{EB} = \left[\underline{T(t_0)}\right]^{EB} \exp\left(\left[\underline{\Omega}^{BE}\right]^{B} \Delta t\right) \tag{3.21}$$

where $\exp(A)$ is the matrix exponential of the matrix A, defined as

$$\exp(A) = \sum_{i=0}^{\infty} \frac{1}{i!} A^i \tag{3.22}$$

Note that the exponential in (3.21) can be solved in closed form; this closed-form solution is related to the 'rotation vector', discussed in more detail in Sect. 8.1.1.

Time Derivative of Euler Angles

We saw also that we can describe a transformation matrix (and therefore a rotation tensor) using three Euler angles (ψ, θ, ϕ). Their differential equations can also be computed, starting from (3.20). Here, we will only state the result[3]:

$$\frac{\mathrm{d}}{\mathrm{d}t}\begin{bmatrix} \phi \\ \theta \\ \psi \end{bmatrix} = \begin{bmatrix} 1 & \sin\phi \tan\theta & \cos\phi \tan\theta \\ 0 & \cos\phi & -\sin\phi \\ 0 & \sin\phi / \cos\theta & \cos\phi / \cos\theta \end{bmatrix} \begin{bmatrix} p \\ q \\ r \end{bmatrix} \tag{3.23}$$

The result is a set of three coupled differential equations. We notice, however, a potential issue: if, for example, $\theta = \pm\pi/2$, then the right-hand side is not defined, and the differential equation breaks down.[4] This singularity manifests itself as a discontinuous jump in the Euler angles, even if the frames are rotating in a continuous fashion—more precisely, for continuous motions the description is not necessarily continuous. This property is highly undesirable, and is the main reason why one should not use Euler angles in implementations. However, if the angles remain small, the Euler angle differential equations may be used successfully, and are very intuitive.

We note from (3.23) that if the roll and pitch angles are small, and the angular velocity is small, then to first order

---

[3] Can you derive it?

[4] This is known as "gimbal lock".

$$\frac{\mathrm{d}}{\mathrm{d}t}\begin{bmatrix}\phi\\ \theta\\ \psi\end{bmatrix} \approx \begin{bmatrix}p\\ q\\ r\end{bmatrix} \tag{3.24}$$

which justifies the convention of calling the p the *roll rate*, even though it is not identical to $\frac{\mathrm{d}}{\mathrm{d}t}\phi$, the *rate of roll* (etc.).

3.3.2 Euler Transformation

We can now rewrite the definition of the rotational time derivative, using the angular velocity, to arrive at the *Euler transformation*. This is an extremely useful and powerful result. Note that the Euler transformation has nothing to do with the Euler angles that we discussed earlier.

Theorem 3.1 *For any two frames A and B, with relative angular velocity* $\underline{\Omega^{BA}}$, *and for a tensor* $\underline{x}$

$$D^A \underline{x} = D^B \underline{x} + \underline{\Omega^{BA}}\, \underline{x} \tag{3.25}$$

A formal proof of this statement is given in Zipfel (2007, Sect. 4.2.4), and from a different perspective in Appendix D of the same.

We can, however, loosely derive this by considering two frames A and B, with their associated coordinate systems, and computing:

$$\left[\underline{D^A x}\right]^A = \left[\frac{\mathrm{d}}{\mathrm{d}t}\underline{x}\right]^A = \frac{\mathrm{d}}{\mathrm{d}t}\left[\underline{x}\right]^A \tag{3.26a}$$

$$= \frac{\mathrm{d}}{\mathrm{d}t}\left([T]^{AB}\left[\underline{x}\right]^B\right) \tag{3.26b}$$

$$= \left[\frac{\mathrm{d}}{\mathrm{d}t}T\right]^{AB}\left[\underline{x}\right]^B + [T]^{AB}\left[\frac{\mathrm{d}}{\mathrm{d}t}\underline{x}\right]^B \tag{3.26c}$$

$$= [T]^{AB}\left[\underline{\Omega^{BA}}\right]^B\left[\underline{x}\right]^B + [T]^{AB}\left[\frac{\mathrm{d}}{\mathrm{d}t}\underline{x}\right]^B \tag{3.26d}$$

$$= [T]^{AB}\left[\underline{\Omega^{BA}}\right]^B\left[\underline{x}\right]^B + [T]^{AB}\left[\underline{D^B x}\right]^B \tag{3.26e}$$

where we used the product rule, and then the transformation matrix differential equation (3.20). We note that we can factor the final result, to have

$$\left[\underline{D^A x}\right]^A = [T]^{AB}\left(\left[\underline{\Omega^{BA}}\right]^B\left[\underline{x}\right]^B + \left[\underline{D^B x}\right]^B\right) \tag{3.27}$$

which matches the expected transformation of the Euler transformation.

This is useful, because it is often natural to take the rotational derivative w.r.t. one frame (e.g. Newton's law requires inertial frames), but then we want to change the frame for coordination. We will use this extensively when we do dynamics, in the next chapter.

Problems for Chapter 3

Problem 3.1 (*3 points*) Referring to Sect. 3.2, prove that it doesn't matter which point A_1 is chosen to define the velocity in (3.10).

Problem 3.2 (*10 points*) You want to film a passing race-car (at point R) with a camera (at point C). You can directly set the angular velocity of the camera with respect to the earth, and want to set it so that the race car appears stationary on the camera feed. We introduce two frames: the E frame is fixed on the ground, and the C frame is fixed to the camera (so that $\underline{1^C}$ points along the barrel of the lens, and $\underline{2^C}$ and $\underline{3^C}$ are perpendiculars). A point P will appear to be stationary on the camera's image if

$$\left[\underline{\frac{\mathrm{d}}{\mathrm{d}t} s_{PC}}\right]^C = \begin{bmatrix} * \\ 0 \\ 0 \end{bmatrix}$$

where $*$ is some number. In other words, the point appears stationary on the images if the second and third components of $\left[\underline{\frac{\mathrm{d}}{\mathrm{d}t} s_{PC}}\right]^C$ are zero (while the value in the first component is irrelevant).

The race-car's displacement w.r.t. the camera, and velocity w.r.t. the earth are as below

$$\left[\underline{s_{RC}}\right]^E = \begin{bmatrix} 10 \\ 0 \\ 0 \end{bmatrix} \mathrm{m}, \quad \left[\underline{v_R^E}\right]^E = \begin{bmatrix} 50 \\ 20 \\ 0 \end{bmatrix} \mathrm{m/s}, \quad [T]^{EC} = \begin{bmatrix} 1 & 0 & 0 \\ 0 & 1 & 0 \\ 0 & 0 & 1 \end{bmatrix}$$

The camera is mounted so that you can only set two components of angular velocity (the yaw rate r and the pitch rate q), while the roll rate is always identically zero, so that

$$\left[\underline{\omega^{CE}}\right]^C = \begin{bmatrix} 0 \\ q \\ r \end{bmatrix} \tag{3.28}$$

Compute the components q and r of the angular velocity $\left[\underline{\omega^{CE}}\right]^C$ so that the race car appears stationary on the image.

Problem 3.3 (*10 points*) For this problem, use the fact from Problem 2.10, for two first-order tensors $\underline{x}$ and $\underline{y}$, that $\underline{K}\left(\underline{x}+\underline{y}\right) = \underline{K}(\underline{x}) + \underline{K}\left(\underline{y}\right)$.

a. Given are three frames A, B, and C. Show that $\underline{\omega}^{AC} = \underline{\omega}^{AB} + \underline{\omega}^{BC}$.
b. Use the result to show that $\underline{\omega}^{AB} = -\underline{\omega}^{BA}$.

Problem 3.4 (*10 points*) A UAV at point B is tracking a car at point C, using a radar sensor that reports the displacement of the car with respect to the UAV in the UAV's coordinate system B as $\left[\underline{s_{CB}}\right]^B$, as well as the rate of change of this quantity $\left[\frac{\mathrm{d}}{\mathrm{d}t}\underline{s_{CB}}\right]^B$. We are interested in the velocity of the car with respect to an earth-fixed reference frame, E, so that we can decide whether to issue a speeding fine. The UAV is not translating with respect to the Earth.

The speed limit is set at 60 miles per hour, or $26.8\,\mathrm{m\,s^{-1}}$, and the UAV's state (from its internal estimator) is known as:

$$\left[\underline{v_B^E}\right]^E = \begin{bmatrix} 0 \\ 0 \\ 0 \end{bmatrix} \mathrm{ms^{-1}}$$

$$[T]^{BE} = \begin{bmatrix} 0.612 & -0.5 & 0.612 \\ 0.354 & 0.866 & 0.354 \\ -0.707 & -0. & 0.707 \end{bmatrix}$$

$$\left[\underline{\omega^{BE}}\right]^B = \begin{bmatrix} 1 \\ 0 \\ 3 \end{bmatrix} \mathrm{rads^{-1}}$$

The measurements from the sensor are:

$$\left[\underline{s_{CB}}\right]^B = \begin{bmatrix} 100 \\ 0 \\ -20 \end{bmatrix} \mathrm{m}; \qquad \left[\frac{\mathrm{d}}{\mathrm{d}t}\underline{s_{CB}}\right]^B = \begin{bmatrix} 8 \\ -300 \\ -18 \end{bmatrix} \mathrm{ms^{-1}}$$

Is the car exceeding the speed limit? Show all work.

Problem 3.5 (*10 points*) A quadcopter (with associated body-fixed frame B) is performing a flip, rotating about its $\underline{2}^B$ axis through 360°. The vehicle started aligned with the Earth-fixed frame, that is $[T]^{BE} = \left[\underline{I}\right]$. It is doing the flip with constant angular velocity w.r.t. the Earth-fixed frame E, with specifically

$$\left[\underline{\omega^{BE}}\right]^B = \begin{bmatrix} 0 \\ 10 \\ 0 \end{bmatrix} \mathrm{rad\,s^{-1}}$$

(a) How long does it take the vehicle to complete a full rotation?
(b) After a quarter of the rotation, the vehicle approaches $\theta = 90°$. Calculate $\left[\frac{\mathrm{d}}{\mathrm{d}t}T\right]^{BE}$ at this time.

(c) What are the time derivatives of the Euler angles when $\theta = 90°$?
(d) How would the solution change if the vehicle were to rotate about its $\underline{1}^B$ axis instead of $\underline{2}^B$ (i.e. with $\left[\underline{\omega}^{BE}\right]^B = (10, 0, 0)\ \text{rad s}^{-1}$)?

Problem 3.6 (*7 points*) An object is rotating about its $\underline{3}^B$ axis relative to an earth-fixed frame, E, at a constant angular velocity of 5 rad s^{-1}. The axis of rotation passes through the object's center of mass, C. The point P is fixed on the object, with displacement to the center of mass:

$$\left[\underline{s}_{PC}\right]^B = (2, 1, 1)\ \text{m}$$

This is illustrated in the figure, below.

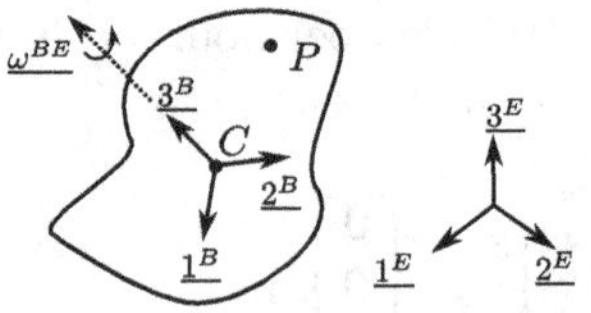

The (instantaneous) transformation matrix is:

$$[T]^{EB} = \begin{bmatrix} 0.5 & 0 & -0.866 \\ 0 & 1 & 0 \\ 0.866 & 0 & 0.5 \end{bmatrix}$$

Compute $\left[\frac{\mathrm{d}}{\mathrm{d}t}\underline{s}_{PC}\right]^E$.

Problem 3.7 (*6 points*) Starting from the definition of the angular velocity tensor:

$$D^B\,\underline{R}^{EB} = \underline{\Omega}^{EB}\,\underline{R}^{EB}$$

and the relationship that

$$\left[\underline{R}^{BA}\right]^B = [T]^{AB}$$

show that the differential equation of the transformation is given by:

$$\left[\frac{\mathrm{d}}{\mathrm{d}t}T\right]^{EB} = [T]^{EB}\left[\underline{\Omega}^{BE}\right]^B.$$

Reference

Zipfel PH (2007) Modeling and simulation of aerospace vehicle dynamics, 2nd edn. AIAA

Chapter 4
Dynamics of Rigid Bodies

The previous chapter gave us the tools to describe *how* things move over time, this chapter tries to answer *why* they move. In this chapter we use the modelling tools that we have developed, and combine them with physical laws to predict how objects move in reaction to external stimuli (forces and torques). We will use two specific laws of nature: Newton's second law, and Euler's law.

This chapter aims to derive generic dynamics expressions. We will use these to understand the mechanism of inertial sensing in Chap. 5, and specialize the dynamics to multicopters in Chap. 6.

4.1 Dynamics of Translation

First, we will relate forces acting on a body to its translational motion. We will require the concept of *translational momentum*, which we can then relate to forces using Newton's second law. First, we define an object's center of mass.

Definition 4.1 Given a body that consists of a collection of points i each with mass m^i at location P_i. The center of mass is that point C for which the following equation holds:

$$\sum_i m^i \, \underline{s_{P_iC}} = \underline{0} \tag{4.1}$$

This is a single, first-order tensor equation wherein the unknown is the position of point C. At first glance, it may seem to have many unknowns (each tensor $\underline{s_{P_iC}}$ is unknown, after all)—this can be resolved by noting that we can generate more

M. W. Mueller, *Dynamics and Control of Autonomous Flight*, Mathematical Engineering,
https://doi.org/10.1007/978-3-031-91871-1_4

equations of the form $\underline{s_{P_i C}} = \underline{s_{P_i P_j}} + \underline{s_{P_j C}}$ where the displacement between P_i and P_j are known.

4.1.1 Translational Momentum

Definition 4.2 Consider a body with mass m^B and center of mass at the point C, and let E be some frame. The translational momentum of the body relative to the frame E is given by $\underline{p_C^E}$, and is defined as

$$\underline{p_C^E} := m^B \, \underline{v_C^E} \tag{4.2}$$

We note that this is a first-order tensor.

4.1.2 Newton's Law of Motion

To express newton's law of motion, we must first introduce the concept of an *inertial frame*.

Definition 4.3 An *inertial frame* is a frame of reference in which any body with zero net force acting upon it is not accelerating.

Loosely speaking, this is a non-accelerating, non-rotating frame. No such thing exists, of course, but depending on the application different approximations are acceptable: we will mostly assume that a frame fixed to the ground may be treated as an inertial frame. For your application, ask yourself how large a force can be approximated well as 'zero net force'. Clearly, if you are flying across the continent, the local ground-fixed frame will be a poor approximation of an inertial frame (the earth rotates a significant amount during a long-haul flight!). Similarly, if you are planning to fly to Mars, you will need an inertial frame approximation which is not attached to the earth (as the earth will move a significant distance around the sun during your mission).

Newton's law then states the following: *In an inertial frame the rate of change of translational momentum equals the force applied.*[1]

We can write this using our notation, expressing the external force as $\underline{f}$, for a body with center of mass C and momentum $\underline{p_C^I}$, in an inertial frame I:

$$D^I \, \underline{p_C^I} = \underline{f} \tag{4.3}$$

[1] This is not a good model of physics if you plan on travelling near the speed of light, even ignoring the complexities of defining an inertial frame.

Note that this is coordinate-free—we needed an inertial frame I, but did not need to specify coordinate systems.

For an object with constant mass m^B (e.g. a battery-operated quadcopter), we can write this as

$$\underline{f} = D^I \underline{p_C^I} = D^I \left(m^B \underline{v_C^I}\right) = m^B \underline{a_C^I}. \tag{4.4}$$

4.2 Attitude Dynamics

We now know how to relate the translational motion of a body to forces acting on it. Next, we look at how a body's orientation evolves as torques act upon it. We will see that attitude dynamics are substantially richer than the translational dynamics.

4.2.1 Inertia Tensor

Definition 4.4 A body B (which we consider a collection of points indexed by i and at positions P_i) with an associated frame B has an inertia tensor referenced to an arbitrary point R defined as

$$\underline{J_R^B} := \sum_i m^i \left(\overline{\underline{s_{P_iR}}}\,\underline{s_{P_iR}}\,\underline{I} - \underline{s_{P_iR}}\,\overline{\underline{s_{P_iR}}}\right) \tag{4.5a}$$

$$= \sum_i m^i\, \overline{\underline{K}\left(\underline{s_{P_iR}}\right)}\,\underline{K}\left(\underline{s_{P_iR}}\right) \tag{4.5b}$$

with $\underline{s_{P_iR}}$, the displacement of point i with respect to the reference point R.

If we coordinate everything in the body-fixed system B, with $\left[\underline{s_{P_iR}}\right]^B = (x_i, y_i, z_i)$, this solves to

$$\left[\underline{J_R^B}\right]^B := \begin{bmatrix} \sum_i m^i \left(y_i^2 + z_i^2\right) & -\sum_i m^i x_i y_i & -\sum_i m^i x_i z_i \\ -\sum_i m^i x_i y_i & \sum_i m^i \left(x_i^2 + z_i^2\right) & -\sum_i m^i y_i z_i \\ -\sum_i m^i x_i z_i & -\sum_i m^i y_i z_i & \sum_i m^i \left(x_i^2 + y_i^2\right) \end{bmatrix} \tag{4.6}$$

- Note that if the body is better modelled as an infinite number of points (a continuum of material), we simply replace the summation with integrals over the density of the material.

- For simple bodies, we can compute this by hand, however it quickly becomes extremely tedious.
- For more complex bodies, we would often use CAD software to compute the mass moment of inertia. Note, however, that some software define the off-diagonal terms differently (specifically, flipping the sign).
- We can observe the mass of an object by measuring its weight, however there is no easy way to measure the mass moment of inertia.
- The inertia tensor is symmetric ($\overline{\underline{J_R^B}} = \underline{J_R^B}$), and it can be shown to be positive definite (that is, for any first-order tensor $\underline{x} \neq 0$ we have $\overline{\underline{x}}\ \underline{J_R^B}\,\underline{x} > 0$).
- Because it is symmetric and positive definite, it has three positive eigenvalues, and three orthogonal eigentensors (called the principal axes of inertia). This means that for any rigid body, there exists a coordinate system fixed to the body in which the inertia tensor is expressed as a diagonal matrix with positive entries.

4.2.2 Angular Momentum

As for translation, we need a concept of momentum before we can do angular dynamics.

Definition 4.5 The angular momentum of a rigid body B with respect to a frame A and referenced to a point R is defined as

$$\underline{l_R^{BA}} = \underline{J_C^B}\ \underline{\omega^{BA}} + m^B\ \underline{K}(\underline{s_{CR}})\ \underline{v_C^A} \tag{4.7}$$

where $\underline{J_C^B}$ is the body's mass moment of inertia about its center of mass (the point C), m^B is the body's mass, and $\underline{K}(\underline{s_{CR}})$ is the skew symmetric form of the displacement of the center of mass C with respect to the reference point R.

Note that, if the reference point R coincides with the center of mass, $\underline{s_{CR}} = \underline{0}$ and $\underline{l_R^{BA}} = \underline{J_C^B}\ \underline{\omega^{BA}}$. If the reference point is not the center of mass, we note that the angular momentum contains the translational momentum in the second term (the product of mass and translational velocity). This former simpler form is what we will typically use when considering objects in free flight.

4.2.3 Moments

Next we require the rotational equivalent of a *force*, which we call the moment of force, or simply 'moment'.

Definition 4.6 A force $\underline{f}$ acting at a point P from some reference point R creates a moment $\underline{n_R}$ defined as

$$\underline{n_R} = \underline{K}(\underline{s_{PR}})\ \underline{f}. \tag{4.8}$$

4.2.4 Euler's Law

We now have all the tools to express the dynamics of attitudes.

In an inertial frame I the rate of change of angular momentum $\underline{l_B^{BI}}$ of a body B with respect to its center of mass C equals the external moment $\underline{n_C}$ applied (where the moment is about the center of mass).

$$D^I \underline{l_C^{BI}} = \underline{n_C} \tag{4.9}$$

If we substitute the angular momentum, we get the below, taking care to note that the derivative is of the *product*:

$$D^I \left(\underline{J_C^B} \; \underline{\omega^{BI}} \right) = \underline{n_C} \,. \tag{4.10}$$

4.3 Coordination of Attitude Dynamics

It is usually most convenient to express the angular dynamics in a body-fixed coordinate system, B. The primary reason for this is that the inertia tensor is constant in the body-fixed coordinate system, for a rigid body.

Here, we will derive the expression for the rate of change of the angular velocity vector composed of the roll, pitch, and yaw rates: $\left[\underline{\frac{\mathrm{d}}{\mathrm{d}t}\omega^{BE}} \right]^B = (\dot{p}, \dot{q}, \dot{r})$ as a function of external moments acting on a body $\underline{n_B}$.

We begin with Euler's law (4.9), and apply Euler's transformation to transform to the rotational derivative in the B frame:

$$\underline{n_B} = D^E \underline{l_B^{BE}} = D^E \left(\underline{J_B^B} \; \underline{\omega^{BE}} \right) \tag{4.11a}$$

$$= D^B \left(\underline{J_B^B} \; \underline{\omega^{BE}} \right) + \underline{\Omega^{BE}} \left(\underline{J_B^B} \; \underline{\omega^{BE}} \right) \tag{4.11b}$$

$$= \underline{J_B^B} \, D^B \, \underline{\omega^{BE}} + \underline{\Omega^{BE}} \; \underline{J_B^B} \; \underline{\omega^{BE}} \tag{4.11c}$$

where we use the fact that $\underline{J_B^B}$ is constant in the B frame, i.e. $D^B \, \underline{J_B^B} = 0$.

Coordinating in B, and rearranging gives

$$\left[\underline{J_B^B} \right]^B \left[\underline{D^B \omega^{BE}} \right]^B = - \left[\underline{\Omega^{BE}} \right]^B \left[\underline{J_B^B} \right]^B \left[\underline{\omega^{BE}} \right]^B + \left[\underline{n_B} \right]^B \tag{4.12}$$

or,

$$\left[\underline{\frac{\mathrm{d}}{\mathrm{d}t}\omega^{BE}} \right]^B = \left(\left[\underline{J_B^B} \right]^B \right)^{-1} \left(- \left[\underline{\Omega^{BE}} \right]^B \left[\underline{J_B^B} \right]^B \left[\underline{\omega^{BE}} \right]^B + \left[\underline{n_B} \right]^B \right) \tag{4.13}$$

where the inverse is guaranteed to exist because the inertia tensor is positive definite.

4.4 Lagrangian Mechanics

The preceding approach to dynamics is typically called 'Newton-Euler', since it encodes Newton's laws in vector format (it was Euler who made Newton's law easy to use). Newton did not originally have the concept of vectors as we understand them, making it much more tedious to use his dynamics.

An alternative approach to dynamics, often used in robotics, is the Lagrangian approach (where we define our degrees of freedom, and derive the equations of motion in these). This approach makes it very easy to deal with constraints (e.g., a link that is fixed to rotate about a hinge); however it is very tedious when dealing with freely moving objects. For a freely moving rigid body, the Newton-Euler approach has only two (relatively simple) tensor equations, while the Lagrangian approach will typically have 6 scalar equations (which will be very ugly, and much harder to gain intuition/insights from). For a nice discussion of Lagrangian mechanics, see e.g. Kasdin and Paley (2011). Note that Lagrangian mechanics just rewrites Newton's law, so that constraints are implicit in the equations (i.e., fundamentally, we're still doing Newton's laws of motion).

Problems for Chapter 4

Problem 4.1 (*12 points*) Compute the mass moment of inertia of the quadcopter shown below: $\left[\underline{J_B^B} \right]^B$ about the quadcopter's center of mass B.

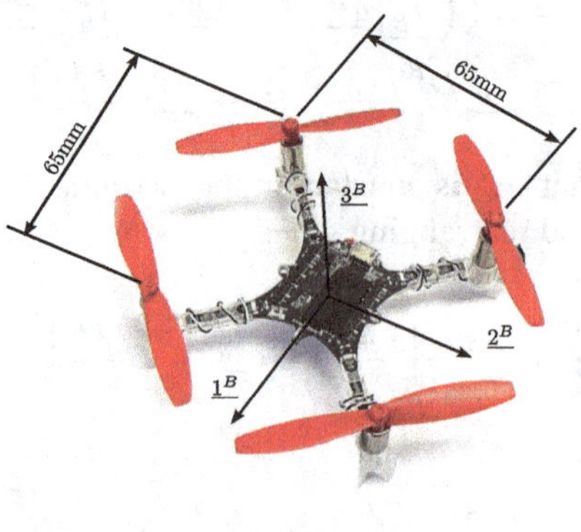

Do this by approximating the vehicle as five particles: 4 particles, one centered at each of the motors, and one further particle at the center of mass. You may assume that the center of mass coincides with the geometric center (the middle of the vehicle). Each particle at the center of a motor has a mass of 3.3×10^{-3} kg; the particle at the center of mass has a mass of 14.5×10^{-3} kg.

Problem 4.2 (*8 points*) You are controlling a space-borne telescope, that is imaging a point on the ground on earth. To get a clean image, the satellite has to maintain a precise (constant) angular velocity, specifically $\left[\underline{\omega^{BI}}\right]^B = (5, 0, 2)$ rad/s. The spacecraft's mass moment of inertia is

$$\left[\underline{J_B^B}\right]^B = \begin{bmatrix} 2 & 1 & 1 \\ 1 & 3 & 0 \\ 1 & 0 & 5 \end{bmatrix} \text{ kg m}^2.$$

Compute the torque that is required to maintain the required angular velocity, $\left[\underline{n_B}\right]^B$.

Problem 4.3 (*12 points*) A spacecraft has four thrusters i located at positions P_i. The magnitude of force produced by each thruster is given by a_i (a scalar), and the thrusters point in the directions $\underline{e_i}$. It is given:

$$\left[\underline{e_1}\right]^B = \begin{bmatrix} 1 \\ 0 \\ 0 \end{bmatrix} \qquad \left[\underline{s_{P_1B}}\right]^B = \begin{bmatrix} 0 \\ 1 \\ 0 \end{bmatrix} \text{ m} \tag{4.14}$$

$$\left[\underline{e_2}\right]^B = \begin{bmatrix} -1 \\ 0 \\ 0 \end{bmatrix} \qquad \left[\underline{s_{P_2B}}\right]^B = \begin{bmatrix} 0 \\ -1 \\ 0 \end{bmatrix} \text{ m} \tag{4.15}$$

$$\left[\underline{e_3}\right]^B = \begin{bmatrix} 0 \\ 1 \\ 0 \end{bmatrix} \qquad \left[\underline{s_{P_3B}}\right]^B = \begin{bmatrix} 1 \\ 0 \\ 0 \end{bmatrix} \text{ m} \tag{4.16}$$

$$\left[\underline{e_4}\right]^B = \begin{bmatrix} 0 \\ -1 \\ 0 \end{bmatrix} \qquad \left[\underline{s_{P_4B}}\right]^B = \begin{bmatrix} -1 \\ 0 \\ 0 \end{bmatrix} \text{ m} \tag{4.17}$$

where B is a body-fixed coordinate system, and B is the spacecraft center of mass.

Compute the moment acting at the vehicle center of mass, $\left[\underline{n_B}\right]^B$ as a function of the values of a_i.

Problem 4.4 (*8 points*) A satellite at point B is orbiting the Earth at a height of 400 km above the earth's surface (the same orbit as the International Space Station). We introduce two frames: E is a convenient inertial frame, and B is a body-fixed frame. The velocity, angular velocity, mass, and mass moment of inertia for satellite are:

$$\left[\underline{v_B^E}\right]^E = \begin{bmatrix} 27.7 \\ 0 \\ 0 \end{bmatrix} \text{ km/s}, \quad \left[\underline{\omega^{BE}}\right]^B = \begin{bmatrix} 0 \\ 4 \\ 5 \end{bmatrix} \text{ rad/s}, \quad m^B = 1000 \text{ kg},$$

$$\left[\underline{J_B^B}\right]^B = \begin{bmatrix} 15 & 0 & 0 \\ 0 & 15 & 0 \\ 0 & 0 & 35 \end{bmatrix} \text{ kg m}^2$$

At the height at which the satellite operates, the acceleration due to gravity is $\left[\underline{g}\right]^E = (0, 0, -8.7)\,\text{m s}^{-2}$.

A thruster on the satellite is located at point R, and produces a thrust $\underline{t}$. The location of thruster and the direction of thrust are as follows:

$$\left[\underline{s_{RB}}\right]^B = \begin{bmatrix} 3 \\ 2 \\ 0 \end{bmatrix} \text{ m}, \quad \left[\underline{t}\right]^B = \begin{bmatrix} -50 \\ 70 \\ 0 \end{bmatrix} \text{ N}$$

Given is that (instantaneously) the two frames align, i.e. $[T]^{BE} = \left[\underline{I}\right]$.

(a) Find the acceleration of the satellite, $\left[\underline{a_B^E}\right]^E$.

(b) Find the angular acceleration of the satellite, $\left[\underline{\frac{\mathrm{d}}{\mathrm{d}t}\omega^{BE}}\right]^B$.

Problem 4.5 (*10 points*) Consider the rocket below, with body-fixed frame B, moving over the Earth with Earth-fixed frame E assumed inertial.

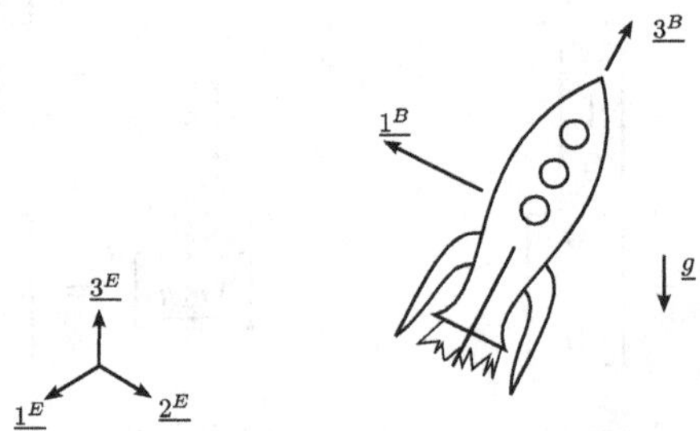

We assume that the mass of the rocket, m^B is constant, and that the rocket is acted upon by two forces: the thrust is given by $\underline{t}$, and the weight, through the acceleration due to gravity $\underline{g}$.

The thrust force is known in the body-fixed coordinate system, $\left[\underline{t}\right]^B$; gravity is known in the earth-fixed coordinate system $\left[\underline{g}\right]^E$, and we know the vehicle's orientation $[T]^{BE}$. Finally, we know the velocity of the rocket with respect to the earth, from ground-based radar: $\left[\underline{v_B^E}\right]^E$.

Provide an equation for the vehicle's acceleration, coordinated in E: $\left[\underline{a_B^E}\right]^E$, in terms of the known quantities.

Problem 4.6 (*8 points*) We consider again the rocket of the Problem 4.5. Let the thrust be given by $\underline{t} = a\,\underline{3^B}$, where a is a scalar. Now, we no longer assume that the mass of the vehicle is constant, and instead we have $\frac{\mathrm{d}}{\mathrm{d}t}m^B = qa$, with q some known proportionality constant.

For the rocket's controller, you need an expression of the rocket's acceleration as a function of the input a. Provide this equation for the vehicle's acceleration, coordinated in E: $\left[\underline{a_B^E}\right]^E$, in terms of the known quantities.

Problem 4.7 (*6 points*) Consider a rigid body consisting of three particles (P_1, P_2, and P_3) rigidly connected w.r.t. a body-fixed frame B. The particles have masses: $m^1 = 1\,\text{kg}$, $m^2 = 10\,\text{kg}$, and $m^3 = 2\,\text{kg}$, respectively. They are arranged so that:

$$\left[\underline{s_{P_2P_1}}\right]^B = \begin{bmatrix}1\\1\\0\end{bmatrix}\text{m;}\quad \left[\underline{s_{P_3P_1}}\right]^B = \begin{bmatrix}2\\0\\0\end{bmatrix}\text{m;} \tag{4.18}$$

Compute the position of the body's center of mass, C, w.r.t. point P_1 as $\left[\underline{s_{CP_1}}\right]^B$.

Problem 4.8 (*6 points*) For the rigid body of the previous problem, compute the body's mass moment of inertia about P_1: $\left[\underline{J_{P_1}^B}\right]^B$.

Problem 4.9 (*5 points*) A gasoline-powered, fixed-wing UAV is flying at a velocity of $\left[\underline{v_B^E}\right]^E = (30, 0, 0)$ m/s relative to a fixed point on the earth, where B is the vehicle's center of mass, and E is an Earth-fixed, inertial frame. Four forces act along the vehicle, given below:

- weight $\left[\underline{f_w}\right]^E = m_B\left(0, 0, -\left\|\underline{g}\right\|\right)$ N
- thrust $\left[\underline{f_t}\right]^E = (20, 0, 10)$ N
- lift $\left[\underline{f_l}\right]^E = (0, 0, 100)$ N
- drag $\left[\underline{f_d}\right]^E = (-15, 0, 0)$ N

The vehicle mass is $m_B = 10\,\text{kg}$, and the vehicle burns gasoline at a rate of $0.1\,\text{kg/s}$. Compute the vehicle's acceleration $\left[\underline{a_B^E}\right]^E$.

Problem 4.10 (*12 points*) A thruster is attached to a brick, which is initially at rest. The brick has a body-fixed frame B, and its center of mass is at point B. Its mass properties are as below (this includes any contributions of the thruster):

$$m^B = 2\,\text{kg} \tag{4.19}$$

$$\left[\underline{I_B^B}\right]^B = \begin{bmatrix} 5 & 0 & 0 \\ 0 & 15 & 0 \\ 0 & 0 & 16 \end{bmatrix} \times 10^{-3}\,\text{kgm}^2 \tag{4.20}$$

The thruster is attached to the brick at point R, and produces a force of 10 N in the direction of $\underline{3^B}$. It is located so that $\left[\underline{s_{RB}}\right]^B = (0.1, 0, 0)$ m.

E is an earth-fixed frame (assumed inertial), and initially the B and E frames coincide so that $[T]^{BE} = \left[\underline{I}\right]$.

(a) Compute the rate of change of the brick's velocity with respect to the earth-fixed frame, coordinated in E, $\left[\frac{\mathrm{d}}{\mathrm{d}t}\underline{v_B^E}\right]^E$.
(b) Compute the angular acceleration of the brick with respect to the earth-fixed frame, coordinated in B, $\left[\frac{\mathrm{d}}{\mathrm{d}t}\underline{\omega^{BE}}\right]^B$.

Problem 4.11 (*8 points*) A fixed-wing UAV is executing a steady turn a steady height and a constant forward speed. A "steady turn" is a maneuver where the angular velocity is constant.

Let B be a body-fixed frame, as shown below, and denote the point corresponding to the vehicle center of mass as B. The vehicle's mass moment of inertia tensor is given below:

$$\left[\underline{J_B^B}\right]^B = \begin{bmatrix} 0.8 & 0 & 0 \\ 0 & 0.7 & 0 \\ 0 & 0 & 1.4 \end{bmatrix} \text{kgm}^2$$

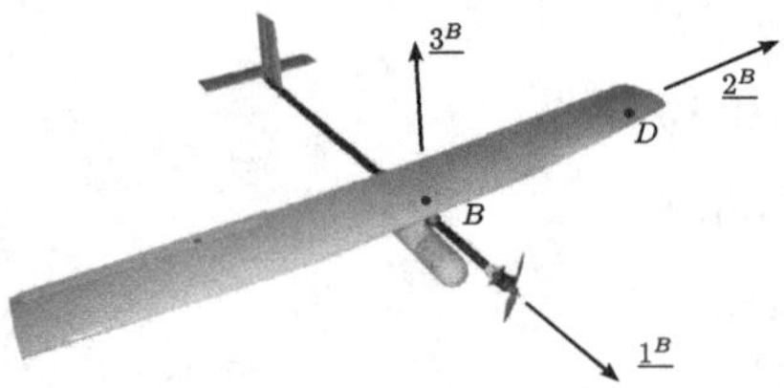

During the steady turn, the vehicle's angular velocity is $\left[\underline{\omega^{BE}}\right]^B = (0, 0.1, 1)$ rad/s.

(a) Compute the moment required during the steady turn.

Suddenly, a mechanical fault causes the UAV's wing tip to bend, and creates a force $\left[\underline{f_f}\right]^B = (0, 0, 0.1)$ N at the position D, where $\left[\underline{s_{DB}}\right]^B = (0, 1, 0)$ m. The fault does not cause a change in the vehicle's center of mass, or mass moment of inertia, nor does it change any other forces/moments acting on the vehicle.

(b) Assuming that all other moments and forces remain the same, what is the angular acceleration of the UAV after the fault?

Problem 4.12 (*4 points*) Using Definition 4.4, show that $\underline{J_R^B}$ is a second-order tensor.

Problem 4.13 (*8 points*) *Changing the reference point:* If we know the mass moment of inertia of a body about its center of mass B as $\underline{J_B^B}$, but want to instead know the inertia about a different point R, we can employ Huygen's theorem. Specifically, let $\underline{s_{RB}}$ be the displacement of R with respect to B, and m^B be the mass of the body. Then

$$\underline{J_R^B} = \underline{J_B^B} + m^B \overline{\underline{s_{RB}}}\, \underline{s_{RB}} \tag{4.21}$$

Prove this using the definition of the inertia tensor and the definition of the center of mass.

Reference

Kasdin NJ, Paley DA (2011) Engineering Dynamics. Princeton University Press

Chapter 5
Sensing and Estimation

An alternative chapter title, 'sense and sensibility', was sadly already taken.

Sensors allow a system to infer information about the world around it. This information may either be purely informative (e.g. to report the air temperature for later analysis) or it may be used for control (e.g. its position relative to the world, so that a robot may avoid collision).

An ideal sensor allows us to measure a quantity of interest directly—e.g., an ideal position sensor would directly measure a position without any errors at every possible time instant. Of course, in reality, no sensor is ideal, and we have to contend with many sources of error; the more we understand about the actual measurement principles of a sensor, the better we will be able to work around its real-world complexity. We will restrict ourselves to *digital* sensors, i.e., where the measurement is available for computation (as opposed to analog sensors). A broad discussion of robotic sensors can be found, for example, in Siegwart et al. (2011).

5.1 Sensor Basics

The most important characteristics of a sensor are the *range, resolution, and measurement frequency*. The *range* is the minimum and maximum values that the sensor reliably can measure; while the *resolution* is the smallest change in the sensed quantity that is reported, typically specified in units of measure per bit. Lastly, the *measurement frequency* is the rate at which the sensor outputs measurements, typically in Hz (that is, number of sensor measurements per second).

M. W. Mueller, *Dynamics and Control of Autonomous Flight*, Mathematical Engineering,
https://doi.org/10.1007/978-3-031-91871-1_5

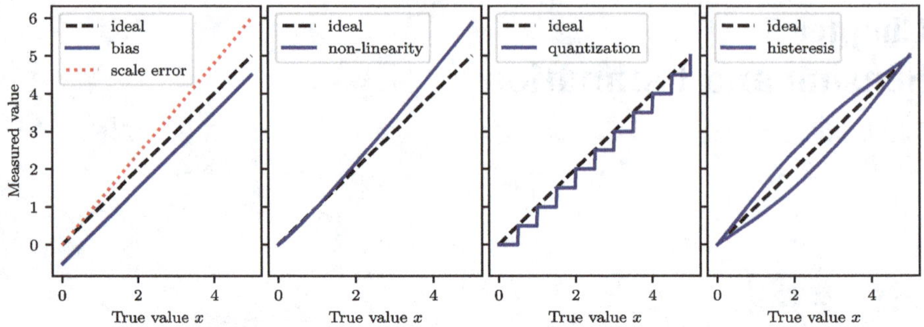

Fig. 5.1 Examples of static sensor imperfections, where the measured values differ from a (constant) true value. From left to right: when the true value is zero, the (non-zero) output of a sensor is its *bias*, and the difference of the measurement slope to identity is called the *scale error*. If the scale error changes over the sensor range, this is called the *non-linearity*; in other words the non-linearity is what remains after correcting for the bias and scale error. *Quantization* is a consequence of the sensor's digital nature—although the true value may be a real number, we can only encode it using finite precision on a computer (note that sensors typically encode with *fixed precision*, which we then convert to `float` for convenient computation.). The last plot shows a sensor whose values differ on whether the true value is increasing or not, known as *histeresis*

For example, a rate gyroscope may have a range of $\pm 2000°/\text{s}$, encoded in 12 bits. The *resolution* of this sensor is $2000/2^{12} \approx 0.5°/\text{s}$ (or, preferably in SI units, $0.0085\,\text{rad}\,\text{s}^{-1}$). The sensor may report measurements at a frequency of 500 Hz.

In addition, you may be interested in a sensor's imperfections, and sources of error. We may distinguish *static* from *dynamic* errors, which are respectively illustrated in Figs. 5.1 and 5.2.

Understanding the physics of a sensor are key for using it well: *how* a sensor measures defines the sources and types of error. The process of characterizing these non-idealities, and correcting for them as best we can, is called *calibration*—the most basic calibration is to correct for bias, and the scale error at zero. It is tempting to treat a sensor as "ideal" after calibration, but it is important to keep in mind that no sensor is perfect, and there will always remain a gap between the numbers that we measure (and thus know perfectly) and the quantity that we want to know. One task of *estimation* will be to mitigate this remaining uncertainty.

5.2 Some Key Sensors

We now model some key flight sensors. The models will allow us to relate the sensor output to underlying quantities of interest, but we will derive *idealized* models, that is we will ignore all the non-idealities discussed above. We will consider first the inertial measurement unit (IMU), specifically the accelerometer and rate gyroscope. Then we will discuss satellite-based localization, barometers, and magnetometers.

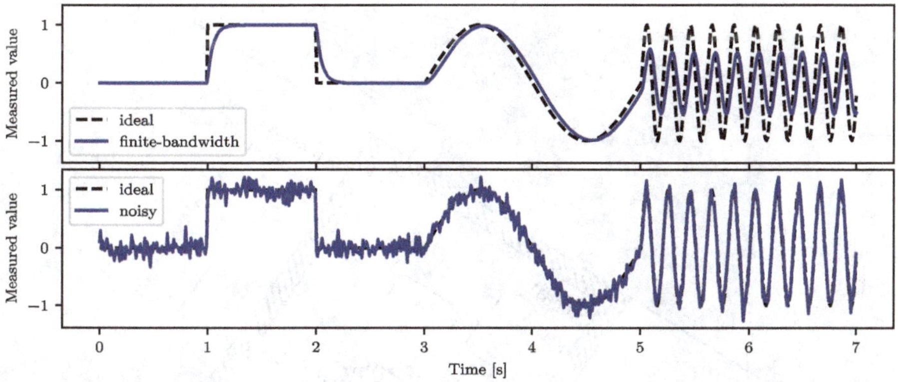

Fig. 5.2 Examples of dynamic sensor imperfections, where the measured values differ from the truth (where the true value changes over time). The top graph shows a *finite bandwidth* sensor, which (in this particular example) cannot faithfully reproduce high-frequency signals—notice that the measurement takes some time to respond to the step change at 1 s; and while it can reproduce the initial low-frequency sine-wave well, it struggles with the later, high-frequency one (showing a notable phase offset, and amplitude change). We would say that this sensor "low-pass filters" the measurement. The bottom graph shows a *noisy* sensor, where the measurement is corrupted by a random noise (where *noise* is an unpredictable error, and the measurement appears 'scattered around the truth')

Beyond this, we will discuss range sensors and optical flow in detail, and briefly discuss approaches using vision, and active approaches like radar and lidar.

5.2.1 Accelerometers

An accelerometer measures *proper acceleration* of an object, that is its acceleration relative to free fall (confusingly, it does not measure an object's acceleration). Conceptually, an accelerometer is a proof-mass, attached to sensors that can measure the forces on the proof mass. The simplest such force sensor is a *strain gauge*, which measures the deformation of a material (and where we can use material properties to relate this deformation to a force). We will assume that the strain gauge material is sufficiently stiff that we can treat it as a rigid connection (and we thus neglect any relative motion/dynamics).

This is illustrated in Fig. 5.3, where an accelerometer is attached to a body B at the point P. We assume the proof mass is constrained with three strain gauges, arranged orthogonally, that measure the force along the measurement directions $\underline{1}^S$, $\underline{2}^S$, and $\underline{3}^S$. Let the corresponding components of force, as measured by the strain gauges in these directions, be ϵ_1, ϵ_2, and ϵ_3.

We note that the total force $\underline{f}$ acting on the proof mass consists of its weight, and the three supports equipped with strain gauges:

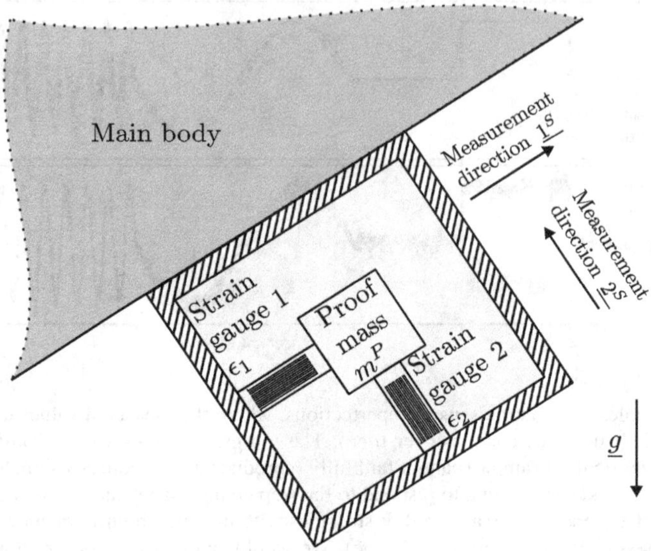

Fig. 5.3 A conceptual accelerometer, showing two strain gauges capable of measuring forces on a proof mass. The third direction (out of the page) is not shown

$$\underline{f} = m^P\, \underline{g} + \epsilon_1\, \underline{1}^S + \epsilon_2\, \underline{2}^S + \epsilon_3\, \underline{3}^S \tag{5.1}$$

Let I be an inertial coordinate system, in which we apply Newton's law to the proof mass. The acceleration of the proof mass, with respect to the inertial $\underline{a}_P^I$, is related to the total force on the proof mass $\underline{f}$, as $m^P\, \underline{a}_P^I = \underline{f}$.

Recall that we know m^P, and measure ϵ_i; we can solve for the known quantities:

$$\frac{\epsilon_1}{m^P}\, \underline{1}^S + \frac{\epsilon_2}{m^P}\, \underline{2}^S + \frac{\epsilon_3}{m^P}\, \underline{3}^S = \underline{a}_P^I - \underline{g} \tag{5.2}$$

The quantity on the right-hand side is known as the *proper acceleration*, the acceleration relative to free fall. Coordinating this in the sensor-fixed coordinate system S gives the *ideal accelerometer*:

$$\left[\,\underline{\alpha}\,\right]^S = \left[\,\underline{a}_P^I\,\right]^S - \left[\,\underline{g}\,\right]^S \tag{5.3}$$

with $\left[\,\underline{\alpha}\,\right]^S = \left(\frac{\epsilon_1}{m^P}, \frac{\epsilon_2}{m^P}, \frac{\epsilon_3}{m^P}\right)$ the output from our strain gauges divided by the proof mass (i.e., these are the numbers our system can measure and report). Thus, the quantity we can measure ($\left[\,\underline{\alpha}\,\right]^S$) is the acceleration of the sensor relative to free-fall, we cannot directly measure the acceleration $\left(\left[\,\underline{a}_P^I\,\right]^S\right)$.

An important note is that we measure this quantity for the point P, i.e., where the sensor is located. In free flight, we are typically interested in the motion of the center of mass of the body, so that ideally the accelerometer is located at the center of mass. If these points differ significantly, then the body's angular velocity will affect the reading (i.e., naïvely treating the accelerometer as though it is located at the center of mass may be very misleading)—see Problem 5.5.

5.2.2 Rate Gyroscopes

Rate gyroscopes measure the angular velocity of the sensor with respect to an inertial frame. Note the difference: an accelerometer measures something related to translational *acceleration*, but a rate gyroscope measures rotational *velocity*.

In very high-end applications, such as nuclear submarines, rate gyroscopes which exploit the finite speed of light are used (look up "Ring Laser Gyroscope"). These can be extraordinarily accurate, but are expensive, large, and heavy.

A much more common technology exploits inertial effects (i.e., forces), and measures angular velocity as it disturbs a mass. To understand this, consider the setup in Fig. 5.4: a wheel W, rotating about a fixed axis $\underline{1^S}$ with respect to a body B. The wheel's angular velocity with respect to the body $\underline{\omega^{WB}}$ is constant with respect to the body, so that $D^B \underline{\omega^{WB}} = 0$, because of a high-bandwidth controller that rejects disturbances. We assume that the sensor can measure torques acting on the spindle around which the wheel rotates, $\underline{n_W}$.

We can use Euler's law to relate the moments acting on the wheel to the angular velocities:

$$D^I \underline{l_W^{WI}} = \underline{n_W} = \underline{J_W^W} D^B \underline{\omega^{WE}} + \underline{\Omega^{BE}} \underline{J_W^W} \underline{\omega^{WE}} \tag{5.4}$$

Expanding $\underline{\omega^{WE}} = \underline{\omega^{WB}} + \underline{\omega^{BE}}$, we get

$$\underline{n_W} = \underline{J_W^W} D^B \underline{\omega^{BE}} + \underline{\Omega^{BE}} \underline{J_W^W} \underline{\omega^{WB}} + \underline{\Omega^{BE}} \underline{J_W^W} \underline{\omega^{BE}} \tag{5.5}$$

If we now assume that the angular velocity of the body is constant, so $D^B \underline{\omega^{BE}} = 0$, and that the angular velocity of the wheel is much larger than that of the body, $\| \underline{\omega^{WB}} \| \gg \| \underline{\omega^{BE}} \|$, we get

$$\underline{n_W} \approx \underline{\Omega^{BE}} \underline{J_W^W} \underline{\omega^{WB}} \tag{5.6}$$

We now coordinate in the body-fixed sensor coordinate system S, wherein $\left[\underline{J_W^W} \right]^S = \operatorname{diag}(J_{xx}, J_{xx}, J_{zz})$, $\left[\underline{\omega^{BE}} \right]^S = (p, q, r)$, and $\left[\underline{\omega^{WB}} \right]^S = (0, 0, \| \underline{\omega^{WB}} \|)$, we have

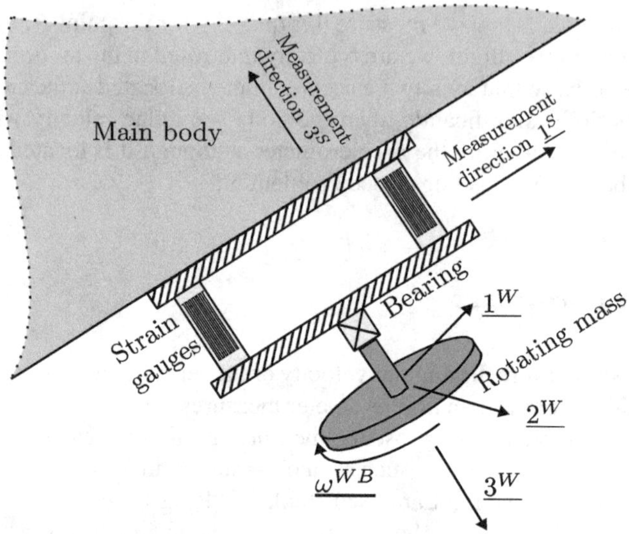

Fig. 5.4 A conceptual rate gyroscope, consisting of a wheel W rotating at constant speed with respect to the body (with body-fixed sensor frame S) about the $\underline{3}^S$ axis. Strain gauges are used to measure the moments that the wheel imparts on the body

$$[\underline{n}_W]^S \approx \begin{bmatrix} 0 & -r & q \\ r & 0 & -p \\ -q & p & 0 \end{bmatrix} \begin{bmatrix} J_{xx} & 0 & 0 \\ 0 & J_{xx} & 0 \\ 0 & 0 & J_{zz} \end{bmatrix} \begin{bmatrix} 0 \\ 0 \\ \| \underline{\omega}^{WB} \| \end{bmatrix} \tag{5.7a}$$

$$= \begin{bmatrix} q \\ -p \\ 0 \end{bmatrix} \| \underline{\omega}^{WB} \| J_{zz} \tag{5.7b}$$

where $\| \underline{\omega}^{WB} \|$ and J_{zz} are known, from the sensor design. Note that the diagonal form of the inertia tensor follows from our assumed symmetry of the wheel; this can be relaxed without causing much difficulty. Thus, a sensor made of one rotating wheel can measure two components of angular velocity; and placing multiple rotating wheels in a non-collinear fashion allows sensing the full angular velocity vector. We will use the symbol γ to refer to the rate gyro output, so that the gyro outputs

$$[\underline{\gamma}]^S = [\underline{\omega}^{BE}]^S \tag{5.8}$$

where S is again a sensor-fixed coordinate system, and most commonly $S = B$.

In a MEMS (short for *Micro Electro-Mechanical System*) rate gyroscope (as is used on almost all UAVs), the sensor will typically have vibrating "forks", rather than a rotating wheel, and the sensor senses deflections rather than moments, but the principle is the same.

5.2.3 Global Navigation Satellite Systems (GNSS)

As the name suggests, a GNSS is a system of satellites that provide global coverage for navigation. The most well-known such system is the Global Positioning System (GPS), but there are other system's such as the European Galileo system, China's BeiDou, or Russia's Glonass. Such a system consists of a number of satellites (called a 'constellation') orbiting the Earth.

Each satellite is equipped with an extremely accurate clock, and periodically broadcasts information about its position and the local time. A user (or aircraft) is equipped with a *receiver* that receives such signals from multiple satellites, and tries to compute the location most consistent with the information received. The receiver, unlike the satellites, does not have an accurate clock. Thus, the receiver is tasked with solving four quantities: its 3D position, and the local time; such a solution is called a 'fix'.

It does this because the transmitted radio signal travels at the speed of light from the satellite to the receiver. Each received signal thus contains the position of the broadcasting satellite, and the time at which the signal was sent. Solving the local position and local time allows the computation of the distance to the satellite, which should also be consistent between the time elapsed between transmission and receipt of the signal. As each satellite thus provides a single scalar piece of information (relating distance the signal travelled to the time elapsed), the receiver needs at least four such measurements to uniquely solve the four unknowns. Where more measurements are available, the receiver can provide more accurate/robust estimates. The above can be modified if we make further assumptions; for example, in a car we may assume that the receiver is constrained onto the surface of the Earth; this removes the need for estimating the height of the receiver.

To get a sense of scale—the radius of the earth is approximately 6 Mm, and the GPS satellites travel on an orbit with radius of approximately 26 Mm. The speed of light is about $300\,\mathrm{Mm\,s^{-1}}$, so that a typical GPS signal travels $<100\,\mathrm{ms}$ to the receiver, and a timing error of $1\,\mu\mathrm{s}$ corresponds to a range error of 30 km. For this reason, we cannot rely on a low-cost clock in the receiver, but instead solve the local time as part of the localization fix. This also makes the satellite's time-keeping crucial, necessitating (e.g.) that their times be corrected for relativistic effects.

It is notable that the quality of a GNSS solution tends to be worse in the vertical direction than in the horizontal directions. An instructive mental model for understanding this is as follows, and is illustrated in Fig. 5.5. Imagine that the receiver has four visible satellites, and can solve the local time exactly using one of these. Next, we can solve our North/South position exactly using a second satellite, so that all that remains is the unknown height, and the unknown East/West position, to be solved from two satellite measurements. With the local time known, the signal propagation time is known for each satellite, and thus the range from each satellite to the receiver is known to within some error. It then uses the remaining two satellites to solve this 2D location, which now lies at the intersection of two rings (one ring for each satellite, centered at the satellite's location when transmitting the message, and with

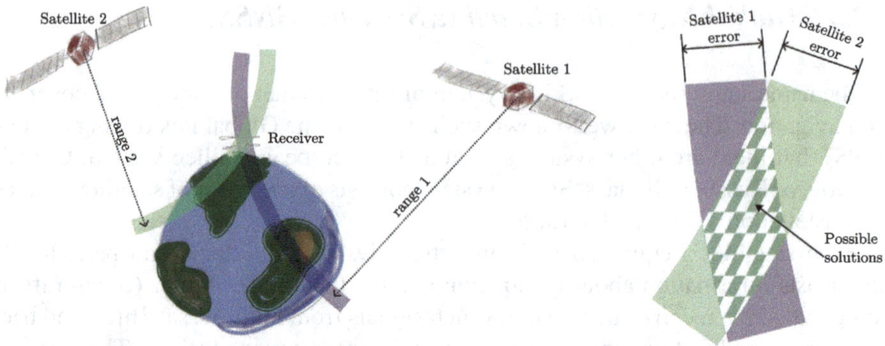

Fig. 5.5 A cartoon illustrating how the satellite constellation geometry affects the quality of the resulting localization. In this simplified model, a receiver on the surface of the earth measures the range to two satellites, shown as the two arcs (each centered on a satellite). Each range is associated with a typical error (e.g., a standard deviation), shown as the width of the arc—the consistent localization solutions are thus in the intersection of the two arcs, shown in detail on the right. Because there is 'more sky' nearer the horizon, this typically leads to worse GNSS height estimates

radius equal to the computed range for that satellite, and width equal to the signal uncertainty). The plausible position solutions then lie at the intersection of these two rings; and the area of that intersection relates to the 'size' of uncertainty.

The lowest uncertainty would occur where the satellites are spaced 90° apart, from the view of the receiver. However, it is much more likely that the satellites are located near the horizon than overhead (there is 'more sky' near the horizon (a large band) than directly overhead (a single point)). Thus, a typical GNSS fix comes from many satellites near the horizon, with few satellites overhead, meaning that the solution has the largest uncertainty/error in the vertical direction. Of course, in reality, the system has to simultaneously solve for the time and 3D position, but the geometric intuition will hold. The jargon for this is *geometric dilution of precision*, or GDOP.

Other sources of error include how the radio signals are bent when travelling through the atmosphere, and where signals are occluded or reflected (e.g., when you're driving through downtown San Francisco). When flying in free air, this is less of a concern, but any application which must operate near large structures must be made safe with the risk that a GNSS fix could be *very wrong*.

There are alternative ways of using GPS data, rather than first solving a fix and then incorporating that fix. One interesting example is provided in Gowda et al. (2016), which uses multiple GPS receivers to reconstruct the orientation of an autonomous aircraft.

5.2.4 Barometer

A barometer measures air pressure. Although this may be useful for its own sake (e.g. in meteorological applications), we are interested in the barometer as a tool to infer height: because air has mass, the pressure decreases the higher the altitude (since the mass of air above decreases).

The atmosphere is complex, and the pressure at a given altitude will depend on many factors, including the temperature at that point. We will assume that the air is stationary relative to the sensor, and that the density of the air is locally constant (this is fine when thinking about moving 100 m up/down, but is a poor assumption for (say) 10 km up/down). We can then straight-forwardly relate the pressure measured to the weight of the atmosphere above the sensor.

Let p be the static pressure, ρ the local density of the air, g the magnitude of the local acceleration due to gravity, and h the height of the sensor. Then

$$\frac{dp}{dh} = \rho g \tag{5.9}$$

or, a change of 1m height at sea level will give a pressure change of approximately 12 Pa. Note how small this is, compared to the absolute pressure (which is around 101 kPa). As an aside, does the standard atmosphere model as given in Appendix A match this? The height calculated from a pressure sensor and an atmospheric model is called the *pressure altitude*.

How good are our assumptions? Is the air locally stationary? Is the ambient pressure constant at a fixed height? Notably, indoor the pressure can vary due to (e.g.) a door opening, so it is worth being cautious when using the barometer.

5.2.5 Magnetometer

A magnetometer measures the local magnetic field vector in its own coordinate system, and thus allows inferring something about the orientation of the vector with respect to the local magnetic field. In an ideal world, such a sensor would be extremely useful—however, in practice magnetometers are difficult to use effectively. We can model this as follows: let $\left[\underline{\mu}\right]^S$ be the output of the magnetometer, in the sensor frame, and $\left[\underline{m}\right]^E$ the known local magnetic field. Then

$$\left[\underline{\mu}\right]^S = [T]^{SE}\left[\underline{m}\right]^E \tag{5.10}$$

A first problem is that the local magnetic field can vary strongly as a function of position, e.g. due to large metal structures (e.g. steel beams) or due to large electric currents. A second issue is that the structure to which the sensor is mounted may

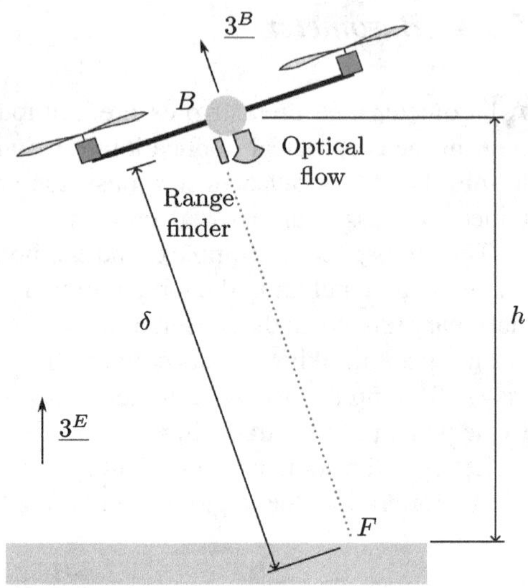

Fig. 5.6 Schematic of the range finder and optical flow sensor model

cause magnetic disturbances, due to deflection of the field, or the generation of local magnetic fields (consider an aircraft with electric motors which work due to magnetic effects). For this reason, the magnetometer is often mounted out-of-the-way, far from the structure of the system, and far from any electric fields (e.g., power cables).

5.2.6 Range Sensors

A range sensor measures the distance from the sensor in a body-fixed direction to the nearest surface. Many different technologies exist to do this, ranging from ultrasound (using the speed of sound), radar or laser time of flight (using the speed of light), or light parallax.

We may model the range finder as measuring a distance δ, as shown in Fig. 5.6, and specifically make the (very strong) assumption that the ground is a flat plane. The range finder is mounted at the center of mass of the vehicle, B, and pointing opposite the vehicle thrust vector $\underline{3^B}$. We assume that the vehicle is flying over a large, horizontal surface, and the point on the ground that the range finder is pointing at is F, so that the displacement of the measured point with respect to the vehicle is $\underline{s_{FB}} = \delta\, \underline{3^B}$.

If we denote the height of the vehicle above ground as h, we can relate this to the range measurement as

$$h = \overline{\underline{3^E}}\, \underline{s_{FB}} = \delta \left(\overline{\underline{3^E}}\, \underline{3^B} \right) \tag{5.11}$$

If we then know the orientation of the vehicle $[T]^{BE}$, we can compute the above as

$$h = \delta \left(\overline{\left[\underline{3}^E \right]}^E \overline{[T]}^{BE} \left[\underline{3}^E \right]^E \right) \tag{5.12}$$

where the term in brackets is equivalent to the bottom-right element of the matrix $\overline{[T]}^{BE}$. Expressed using Euler angles, we get

$$h = \delta \cos \phi \cos \theta \tag{5.13}$$

which matches our intuition—the roll and pitch angles matter, but not the yaw angle.

5.2.7 *Optical Flow*

An optical flow sensor uses successive image frames to estimate the image motion from one frame to the next. It is, in essence, a camera which tracks the apparent motion ('optical flow') at the center of its image. These sensors are used to measure the motion of optical computer mice, for example.

The detailed principle of operation of these sensors is not of particular interest, but we are interested in how the measured quantity relates to the state of our system; we thus again derive the idealized sensor model.

Referring again to Fig. 5.6: let F be a point, instantaneously at the center of the flow sensor's field of view, but fixed to the ground; and let B be the point at the center of the sensor. The sensor measures two scalar quantities, σ_1 and σ_2, which have units rad s^{-1}, and are related to the apparent motion of the point F, from the perspective of the sensor (and thus, from the perspective of the vehicle). The sensor is aligned so that σ_1 aligns with the body-fixed $\underline{1}^B$ direction, and σ_2 with $\underline{2}^B$. Note that, on a two-dimensional image, there can be no apparent motion in the $\underline{3}^B$ direction.

We denote the relative position of the point with respect to the body as $\underline{s_{FB}}$, so that the apparent motion is:

$$\begin{bmatrix} \sigma_1 \\ \sigma_2 \end{bmatrix} = \begin{bmatrix} 1 & 0 & 0 \\ 0 & 1 & 0 \end{bmatrix} \frac{1}{\| \underline{s_{FB}} \|} \left[\underline{\frac{\mathrm{d}}{\mathrm{d}t} s_{FB}} \right]^B \tag{5.14}$$

We now simply need to manipulate the last element:

$$\begin{aligned}
\left[\underline{\frac{\mathrm{d}}{\mathrm{d}t} s_{FB}} \right]^B &= \left[\underline{D^B s_{FB}} \right]^B && (5.15a) \\
&= \left[\underline{D^E s_{FB}} \right]^B + \left[\underline{\Omega^{EB}} \right]^B \left[\underline{s_{FB}} \right]^B && (5.15b) \\
&= - \left[\underline{D^E s_{BF}} \right]^B - \left[\underline{\Omega^{BE}} \right]^B \left(-\delta \left[\underline{3}^B \right]^B \right) && (5.15c) \\
&= - \left[\underline{v_B^E} \right]^B + \| \underline{s_{FB}} \| \left[\underline{\Omega^{BE}} \right]^B \left[\underline{3}^B \right]^B && (5.15d)
\end{aligned}$$

where we substituted the range sensor output.

Thus, the optical flow sensor output is:

$$\begin{bmatrix}\sigma_1\\ \sigma_2\end{bmatrix} = \begin{bmatrix}1&0&0\\0&1&0\end{bmatrix}\left(-\frac{1}{\|\underline{s_{FB}}\|}\left[\underline{v_B^E}\right]^B + \left[\underline{\Omega^{BE}}\right]^B\left[\underline{3^B}\right]^B\right) \tag{5.16}$$

If we know the angular velocity of the vehicle $\left[\underline{\omega^{BE}}\right]^B$ (perhaps from a rate gyroscope), we know the distance to the measured point (e.g., from a range finder), we can relate the two components of optical flow to the vehicle's translational velocity.

5.2.8 *Cameras, Radar, and Lidar*

In addition to the above, an autonomous aircraft may be equipped with many other sensors. Most typical, are cameras: these are low-cost, low-power, and provide a lot of data. Moreover, vision is the primary sensor used by human pilots, so it is clear that they will be useful. A typical approach to using cameras is to take the stream of images from a camera, and extract *visual features*. These are parts of the image with high visual contrast, i.e., parts of the image which are easy to detect. These visual features are then tracked over multiple frames, from which it is possible to reconstruct both the 3D position of the feature, as well as the motion of the camera (this process is called *structure from motion*). Conceptually, each of these visual features behaves somewhat like the single point we imagine the optical flow sensor to track. Of course, cameras are useful for other things as well—they may be used for object detection (e.g., is there someone else on the runway I'm going to land at?).

Radar is widely used in aviation, and comes in many varieties with many complexities. Loosely, radar uses radio waves (i.e., longer wavelengths than visible light), and looks to see reflections of a transmitted radio signal. This is extremely useful for detecting other objects in the sky. Usefully, radar typically returns both the range, as well as the closing speed, of detected objects (with the speed being measurable due to Doppler shift). Radar tends to have low resolution.

Similar to radar, there is lidar, which uses emitted light to create high-resolution 3D maps. This is typically done by sweeping a laser beam through space (typically at infrared light frequencies). Such sensors give very high resolution data, and are especially useful for creating detailed 3D maps.

5.3 Estimation

An estimator reconstructs the state of the system, given the signals from the sensors, and knowledge of the system dynamics. We will discuss here a cascaded approach to state estimation, and we will do it in increasing levels of complexity:

- estimating tilt given measurements from the accelerometer and rate gyroscope;
- estimating height, relative speed, height, and tilt given measurements from the accelerometer, rate gyroscope, and range sensor; and
- estimating the full dynamic state.

The fundamental idea is that we will have some mechanism for *predicting* the change in state; then we'll have a separate mechanism for *measuring* the state. We will do this in a very *ad-hoc* fashion,[1] and build estimators from intuition, for each particular problem. The results will be 'good enough', but not necessarily 'good' (nevermind 'great'). There is a rich literature on systematic approaches to state estimation, with especially the *Kalman filter* being very widespread in aerospace. We discuss these more in Sect. 8.3.

5.3.1 Tilt Estimation

The *tilt* of an object is defined as the two rotational degrees of freedom that determine the orientation with respect to gravity (that is, the roll and pitch angles), but neglecting the rotation around the gravity vector (that is, the yaw angle). We will use the measurements from the accelerometer, enforcing the (restrictive) assumption that the vehicle is on average not accelerating in translation.

Note that this is a somewhat circular argument in-flight, and is begging the question—we need to estimate the tilt for feedback control, so that we can control the vehicle to hover, so that the vehicle does not accelerate in translation. However, for estimating the vehicle's tilt, we are assuming that the vehicle is not accelerating. We will come back to this issue later, and reason about why it is often allowable in practice.

For the estimator, we will immediately use Euler angles. This is because it is most simple for our derivation, but it is *not the approach to use in any serious application*. Instead, one would use very similar ideas, but an orientation encoding that does not have singularities (and the other drawbacks) of Euler angles. Some discussion on how to use orientations in state estimation is provided in Mueller et al. (2017).

No Translational Motion

Consider a quadcopter that is pinned on a spherical joint at its center of mass, so that it is free to rotate, but cannot translate. We assume that the accelerometer is located at the center of mass, and are interested in using the rate gyroscope and accelerometer to estimate the vehicle's tilt (the rate gyroscope will *predict*, and the accelerometer will *measure* something about the tilt).

Since the vehicle is pinned, its acceleration is identically zero, $\underline{a}_B^E = 0$. In this case, the accelerometer therefore measures only negative gravity: $\underline{\alpha} = -\underline{g}$. Coordinating in the body-fixed system, and substituting the transformation-matrix in terms of the yaw-pitch-roll angles gives

[1] A less charitable description would be that we are constructing a 'terrible hack'.

$$\left[\underline{\alpha}\right]^B = -[T]^{BE}\left[\underline{g}\right]^E + \left(\nu_{\alpha_1}, \nu_{\alpha_2}, \nu_{\alpha_3}\right) \tag{5.17}$$

where ν_{α_i} are the noises corrupting the measurement. We rewrite this, using the shorthand cx for $\cos x$ and sx for $\sin x$, as

$$\begin{bmatrix}\alpha_1\\ \alpha_2\\ \alpha_3\end{bmatrix} = -\begin{bmatrix}\mathrm{c}\psi\mathrm{c}\theta & \mathrm{s}\psi\mathrm{c}\theta & -\mathrm{s}\theta\\ \mathrm{c}\psi\mathrm{s}\theta\mathrm{s}\phi - \mathrm{s}\psi\mathrm{c}\phi & \mathrm{s}\psi\mathrm{s}\theta\mathrm{s}\phi + \mathrm{c}\psi\mathrm{c}\phi & \mathrm{c}\theta\mathrm{s}\phi\\ \mathrm{c}\psi\mathrm{s}\theta\mathrm{c}\phi - \mathrm{s}\psi\mathrm{s}\phi & \mathrm{s}\psi\mathrm{s}\theta\mathrm{c}\phi - \mathrm{c}\psi\mathrm{s}\phi & \mathrm{c}\theta\mathrm{c}\phi\end{bmatrix}\begin{bmatrix}0\\ 0\\ -\left\|\underline{g}\right\|\end{bmatrix} + \begin{bmatrix}\nu_{\alpha_1}\\ \nu_{\alpha_2}\\ \nu_{\alpha_3}\end{bmatrix} \tag{5.18a}$$

$$= \begin{bmatrix}-\sin\theta\\ \cos\theta\sin\phi\\ \cos\theta\cos\phi\end{bmatrix}\left\|\underline{g}\right\| + \begin{bmatrix}\nu_{\alpha_1}\\ \nu_{\alpha_2}\\ \nu_{\alpha_3}\end{bmatrix} \tag{5.18b}$$

$$\approx \begin{bmatrix}-\theta\\ \phi\\ 1\end{bmatrix}\left\|\underline{g}\right\| + \begin{bmatrix}\nu_{\alpha_1}\\ \nu_{\alpha_2}\\ \nu_{\alpha_3}\end{bmatrix} \tag{5.18c}$$

Where we additionally make the simplifying assumption that the angles are small.

As we had intuitively argued, the accelerometer only gives us information on two directions of the attitude (the roll and pitch angles). Though one could, in principle, directly use the measurements α_1 and α_2 to determine θ and ϕ, they are likely to be quite noisy, and will therefore lead to noisy state estimates.

Instead, we will combine these measurements with the rate gyroscope measurements, γ_i, using the small angle approximation so that:

$$\begin{bmatrix}\gamma_1\\ \gamma_2\\ \gamma_3\end{bmatrix} = \begin{bmatrix}p\\ q\\ r\end{bmatrix} + \begin{bmatrix}\nu_{\gamma_1}\\ \nu_{\gamma_2}\\ \nu_{\gamma_3}\end{bmatrix} \tag{5.19a}$$

$$\begin{bmatrix}\dot\phi\\ \dot\theta\\ \dot\psi\end{bmatrix} = \begin{bmatrix}1 & \sin\phi\tan\theta & \cos\phi\tan\theta\\ 0 & \cos\phi & -\sin\phi\\ 0 & \sin\phi/\cos\theta & \cos\phi/\cos\theta\end{bmatrix}\begin{bmatrix}p\\ q\\ r\end{bmatrix} \tag{5.19b}$$

$$\approx \begin{bmatrix}1 & 0 & \theta\\ 0 & 1 & -\phi\\ 0 & \phi & 1\end{bmatrix}\left(\begin{bmatrix}\gamma_1\\ \gamma_2\\ \gamma_3\end{bmatrix} - \begin{bmatrix}\nu_{\gamma_1}\\ \nu_{\gamma_2}\\ \nu_{\gamma_3}\end{bmatrix}\right) \tag{5.19c}$$

$$\approx \begin{bmatrix}\gamma_1\\ \gamma_2\\ \gamma_3\end{bmatrix} - \begin{bmatrix}\nu_{\gamma_1}\\ \nu_{\gamma_2}\\ \nu_{\gamma_3}\end{bmatrix} \tag{5.19d}$$

We thus see that the accelerometer gives us information about the angles' values, while the rate gyroscope gives us information about their rate of change. We thus propose the following simple estimation strategy, where we use an overhat to indicate an estimate, i.e. $\hat{x}$ indicates our estimate of the quantity x. We will immediately write the estimator in discrete time, using the fact that the sensors provide measurements at discrete times, spaced a time Δt apart. We thus use the notation $\hat{\theta}(k)$ to indicate the estimate of $\theta(k\Delta t)$.

If our accelerometers were noise-free (i.e. $\nu_{\alpha_i} = 0$), we could use the following simple rule:

$$\hat{\theta}(k) = -\left\| \underline{g} \right\|^{-1} \alpha_1(k). \tag{5.20}$$

If, instead, our rate gyroscopes were noise-free (i.e. $\nu_{\gamma_i} = 0$), and we knew exactly the orientation at time $k = 0$, we could just integrate the rate gyroscope measurements:

$$\hat{\theta}(k) = \hat{\theta}(k-1) + \left(\gamma_2(k) - \hat{\phi}(k-1)\gamma_3(k)\right) \Delta t. \tag{5.21}$$

Of course, neither the accelerometer nor the rate gyroscope are noise-free, nor do we know the angles exactly at time zero. What we then do, is we take a weighted average of these two estimators, which allows us to trade off the noise in the sensors, and our initial uncertainty. We introduce the scalar weight $\rho \in [0, 1]$, and use

$$\hat{\phi}(k) = (1-\rho)\left(\hat{\phi}(k-1) + \left(\gamma_1(k) + \hat{\theta}(k-1)\gamma_3(k)\right)\Delta t\right) + \rho\left(\left\| \underline{g} \right\|^{-1} \alpha_2(k)\right) \tag{5.22}$$

$$\hat{\theta}(k) = (1-\rho)\left(\hat{\theta}(k-1) + \left(\gamma_2(k) - \hat{\phi}(k-1)\gamma_3(k)\right)\Delta t\right) + \rho\left(-\left\| \underline{g} \right\|^{-1} \alpha_1(k)\right) \tag{5.23}$$

We note that, if $\rho = 0$, we rely solely on the rate gyroscope, and if $\rho = 1$, we rely solely on the accelerometer. An optimal value for ρ can be computed using tools from stochastic state estimation, specifically the Kalman filter. Here, we will simply treat it as a tuning factor – it's easy to search over potential values, since ρ is scalar, and typically we use $\rho \ll 1$.

The third orientation degree of freedom, the yaw angle, is of course not estimated by the above. This may be remedied by using a magnetometer (if one is confident in the ambient magnetic field), or for very local, short-term operations, by simply integrating the rate gyroscope for yaw.

Non-zero Translational Motion

In realistic situations, where we do not pin the quadcopter down, we may still be interested in using this kind of estimator. We will now discuss situations where this is valid.

Let $\underline{a_B^E}$ be the (unknown) acceleration profile, and then the accelerometer measures: $(\alpha_1, \alpha_2, \alpha_3) = \left[\underline{a_B^E}\right]^B - \left[\underline{g}\right]^B + \left(\nu_{\alpha_1}, \nu_{\alpha_2}, \nu_{\alpha_3}\right)$. We immediately see the problem: there are three degrees of freedom in the unknown acceleration $\left[\underline{a_B^E}\right]^B$, meaning that we can make the accelerometer output arbitrary measurements at any orientation (and, therefore, without additional assumptions we cannot reconstruct the tilt angles from the accelerometer measurements).

However, if we assume that the acceleration is zero on average, we can still employ the estimator of (5.22) and (5.23), though we may have to adapt the constant ρ. To make the reasoning explicit, let us assume $\mathbb{E}\left[\left[\underline{a_B^E}\right]^E\right] = 0$, or that the expected acceleration vector, expressed in the earth-fixed coordinate system, is zero. We use the notation $\mathbb{E}[x]$ to denote the expected value of some quantity x. Usually it is OK to think of the expectation of x as its "average".[2]

Taking now expectation of the accelerometer outputs, we have:

$$\mathbb{E}[(\alpha_1, \alpha_2, \alpha_3)] = \mathbb{E}\left[\left[\underline{a_B^E}\right]^B - \left[\underline{g}\right]^B + \left(\nu_{\alpha_1}, \nu_{\alpha_2}, \nu_{\alpha_3}\right)\right] \tag{5.24a}$$

$$= \mathbb{E}\left[[T]^{BE}\left[\underline{a_B^E}\right]^E - \left[\underline{g}\right]^B + \left(\nu_{\alpha_1}, \nu_{\alpha_2}, \nu_{\alpha_3}\right)\right] \tag{5.24b}$$

If we now apply linearity of expectation, and assume $\mathbb{E}\left[\nu_{\alpha_i}\right] = 0$, we get

$$\mathbb{E}[(\alpha_1, \alpha_2, \alpha_3)] = \mathbb{E}\left[-\left[\underline{g}\right]^B\right] = -\mathbb{E}\left[[T]^{BE}\left[\underline{g}\right]^E\right] \tag{5.25}$$

We now see that this is very similar to the equation that we used to derive the estimator for zero acceleration, Eq. (5.17), which motivates using the same algorithm (5.22) and (5.23).

Note, however: this is only valid if we know (somehow) that $\mathbb{E}\left[\underline{a_B^E}\right] = 0$. Specifically, note that the estimator is not "aware" of this assumption, and will not detect that the quadcopter is violating the assumption (until, of course, you hit the ground). One way you may know that $\mathbb{E}\left[\underline{a_B^E}\right] = 0$ is if you know that there is an external entity (e.g. a pilot) that is controlling the vehicle so that it is hovering; or by looking at position measurements (e.g., from GPS).

5.3.2 *Estimating the Tilt and Height*

Now that the system is capable of estimating its tilt (from the previous section) we add the downward-facing rangefinder of Sect. 5.2.6 to estimate the height and vertical velocity. We refer to Fig. 5.6 for the points and directions to be used. The

[2] One problem is that one-off events don't have an "average", but they do have an "expectation". For example, what will the temperature be at noon next Saturday? This will happen only once, so it doesn't make sense to talk about 'average'; however, it is an uncertain thing, so it does make sense to talk of expectation.

below estimator is best seen as a “terrible hack”, which will get the job done and be fairly easy to understand; but is in every reasonable sense the wrong way to do it. The ‘right way’ (or at least, a more right way) is more complicated, and is left for Sect. 8.3.

We derived a relationship between vehicle height h and the range sensor output δ in (5.12). Substituting now our estimated attitude $\hat{\theta}$ and $\hat{\phi}$ we can generate a pseudo-measurement of height, h_{meas} as

$$h_{\text{meas}}(k) = \delta_{\text{meas}}(k) \cos \hat{\theta}(k) \cos \hat{\phi}(k) \tag{5.26}$$

Using these height measurements, we define a vertical velocity pseudo-measurement $v_{3,\text{meas}}$ by taking finite differences over two sequential height pseudo-measurements:

$$v_{3,\text{meas}}(k) = \frac{1}{\Delta t} \left(h_{\text{meas}}(k) - h_{\text{meas}}(k-1) \right) \tag{5.27}$$

We then fuse these into measurements with our typical predictor-corrector approach: choose some mixing constants ρ_h and ρ_v, then estimate height and vertical velocity as:

$$\hat{h}(k) = (1-\rho_h)\left(\hat{h}(k-1) + \hat{v}_3(k-1)\Delta t\right) + \rho_h h_{\text{meas}}(k) \tag{5.28}$$

$$\hat{v}_3(k) = (1-\rho_v)\left(\hat{v}_3(k-1) + \hat{a}_3(k-1)\Delta t\right) + \rho_v v_{3,\text{meas}}(k) \tag{5.29}$$

With $\hat{a}_3$ being our best guess of vertical acceleration (at its simplest, just set to zero).

5.3.3 Estimating the Tilt, Height, and Velocity

The system now has an estimation pipeline that consumes measurements from the accelerometer, rate gyroscope, and range finder; and estimates the vehicle angular velocity, tilt angles, and vertical state. In this last step, we will process optical flow measurements to estimate the vehicle’s lateral velocity as well.

Referring to the optical flow model derived in (5.16), we make the following simplifying assumptions:

- The range sensor directly measures the distance to the point the optical flow sensor measures, $\| \underline{s_{FB}} \| = \delta$
- The body angular velocity is measured by the rate gyroscope, $\left[\underline{\omega^{BE}}\right]^B = \left[\underline{\gamma}\right]^B$
- The orientation of the body is close enough to identity $[T]^{BE} \approx \left[\underline{I}\right]$ (note that this is a super dubious (and lazy) assumption!)

We can then solve (5.16) to get pseudo-measurements for the horizontal components of velocity:

$$v_{1,\text{meas}} = \delta(k)\left(-\sigma_1(k) + q(k)\right) \tag{5.30}$$

$$v_{2,\text{meas}} = \delta(k)\left(-\sigma_2(k) - p(k)\right) \tag{5.31}$$

We can then fuse these into velocity estimates $\hat{v}_i$ using a predictor-corrector structure:

$$\hat{v}_i(k) = (1 - \rho_v)\left(\hat{v}_i(k-1) + \hat{a}_i(k-1)\Delta t\right) + \rho_v v_{i,\text{meas}}(k) \tag{5.32}$$

with $\hat{a}_i$ our best guess of horizontal acceleration, which we can at our laziest simply set to zero.

The end result of the above is a state estimator that uses the sensors sequentially, and can estimate enough of the system state for hover control. As we have no sensors for horizontal position, we do not estimate these.

5.3.4 Estimating Position with GPS

Incorporating GPS sensors into our hierarchical state estimation strategy is now straight-forward. We have an estimated 3D velocity from the preceding, which we can use for the prediction phase of 3D position estimation. The correction is then done with the GPS measurement of 3D position.

5.3.5 Visual-Inertial Odometry (VIO)

Visual-inertial odometry is an integrated approach to local state estimation that uses a camera and IMU (accelerometer and rate gyroscope). It is purely *local*, as it estimates the position and orientation relative to an arbitrary initial condition, unlike (say) GPS which gives position relative to a specified position. Many off the shelf solutions exist, a good example is OpenVINS by Geneva et al. (2020).

Problems for Chapter 5

Problem 5.1 (*10 points*) An object with center of mass C is sliding down a ramp, at constant velocity and orientation relative to the ground. The object's frame B is oriented relative to the Earth-fixed frame E as

$$[T]^{BE} = \begin{bmatrix} 1 & 0 & 0 \\ 0 & \sqrt{1/2} & \sqrt{1/2} \\ 0 & -\sqrt{1/2} & \sqrt{1/2} \end{bmatrix}$$

The object has an accelerometer and a rate gyroscope, with axes aligned with B. Compute the output of both the accelerometer and rate gyroscope.

Problem 5.2 (*7 points*) A quadcopter of mass 1 kg is flying through a complex wind field, and is being acted by a disturbance force $\underline{f_w}$ due to the wind in addition to the usual multicopter forces. The frame B is fixed to the vehicle, and E is a ground-fixed, inertial, frame. The vehicle is equipped with an accelerometer, mounted at its center of mass, which measures the output $\left[\underline{\alpha}\right]^B$. Given is the following:

$$[T]^{EB} = \begin{bmatrix} 0.6 & -0.8 & 0 \\ 0.8 & 0.6 & 0 \\ 0 & 0 & 1 \end{bmatrix}, \quad c_1 = c_2 = 3\text{N}, c_3 = c_4 = 4\text{N} \quad \left[\underline{\alpha}\right]^B = \begin{bmatrix} 3 \\ 4 \\ 10 \end{bmatrix} \text{m/s}^2$$

where c_i are the individual propeller force magnitudes.

What is the disturbance force, $\left[\underline{f_w}\right]^E$?

Problem 5.3 (*8 points*) Your phone is in your pocket, while you are sitting in a stationary car. At time $t = t_0$ car is at rest, and your phone's accelerometer measures $\left[\underline{\alpha(t_0)}\right]^S = (7.85, -5.9, 0)\,\text{m/s}^2$, where S is a coordinate system attached to the sensor, and (t_0) is the time index.

The car accelerates horizontally, and the accelerometer now measures

$$\left[\underline{\alpha(t_1)}\right]^S = (10.85, -5.9, 4)\,\text{m/s}^2$$

at time t_1. The phone's rate gyroscope measured $\left[\underline{\gamma(t)}\right]^S = (0, 0, 0)\,\text{rad s}^{-1}$ from t_0 to t_1. Compute the magnitude of the car's acceleration at time t_1.

Problem 5.4 (*8 points*) A quadcopter is sitting stationary on a tilted platform, prior to taking off. The vehicle is aligned with the Earth-fixed frame so that the yaw angle is $\psi = 0$.

The accelerometer on the vehicle measures $\left[\underline{\alpha}\right]^B = (2, 4, 8.73)\,\text{m s}^{-2}$. What are the vehicle's roll and pitch angles?

Problem 5.5 (*8 points*) Consider a rigid body B, with two accelerometers attached. The first is attached at the center of mass, point C, and gives measurement $\left[\underline{\alpha_C}\right]^B$. The second is attached at a different point P, and measures $\left[\underline{\alpha_P}\right]^B$. Provide an expression for $\left[\underline{\alpha_P}\right]^B$ in terms of $\left[\underline{\alpha_C}\right]^B$, the (known and constant) displacement of the points $\left[\underline{s_{PC}}\right]^B$, the angular velocity of the body $\left[\underline{\omega^{BI}}\right]^B$, and angular acceleration $\left[\underline{\frac{\mathrm{d}}{\mathrm{d}t}\omega^{BI}}\right]^B$.

Problem 5.6 (*5 points*) A quadcopter (in the "standard" configuration) commands a total thrust force $c_\Sigma = 5\text{N}$. In addition to the thrust force, the vehicle is only subject the force of gravity.

The vehicle has a total mass $m^B = 0.4\,\text{kg}$. Let B be the body-fixed coordinate system, and E the usual inertial coordinate system.

Compute the measurement of a perfect accelerometer, located at the vehicle's center of mass, coordinated in the body-fixed system.

Relative to the quadcopter's starting position S, the center of mass B is at a location $\left[\underline{s_{BS}}\right]^E = (2, 0, 10)$ m, and the quadcopter is oriented so that the Euler yaw/pitch/roll angles are $\psi = \theta = \phi = 0°$; the vehicle's angular velocity is $\left[\underline{\omega^{BE}}\right]^B = (0, 0, 0)$ rad/s.

Problem 5.7 (*6 points*) Your spacecraft's rate gyroscope has failed, and you are tasked with using its star trackers to determine your angular velocity. A star tracker measures the direction (in the body-fixed B coordinate system) of known stars. Two stars of interest are being tracked, with the unit tensors $\underline{a}$ and $\underline{b}$ pointing from the vehicle center of mass to each of star, respectively. These stars are effectively infinitely far from you, and are fixed with respect to the galactic frame, G. Given is

$$\left[\underline{a}\right]^B = \begin{bmatrix} 1 \\ 0 \\ 0 \end{bmatrix}, \quad \left[\underline{b}\right]^B = \begin{bmatrix} 0.6 \\ 0 \\ 0.8 \end{bmatrix}, \quad \left[\underline{\frac{\mathrm{d}}{\mathrm{d}t} a}\right]^B = \begin{bmatrix} 0 \\ -3 \\ 1 \end{bmatrix} \mathrm{s}^{-1},$$

$$\left[\underline{\frac{\mathrm{d}}{\mathrm{d}t} b}\right]^B = \begin{bmatrix} -0.8 \\ -3.4 \\ 0.6 \end{bmatrix} \mathrm{s}^{-1}$$

Compute $\left[\underline{\omega^{BG}}\right]^B$.

Problem 5.8 (*10 points*) An object, which is carried by a person, is moving horizontally (that is, with no vertical velocity) and with zero acceleration ($\underline{a_B^E} = \underline{0}$) over a flat horizontal surface. The object is equipped with sensors, whose output is given in the body-fixed frame B. The optical flow and range sensors point along $-\underline{3^B}$ and see a feature at the point F on the ground (as they did for the quadcopter we analyzed). The earth-fixed E frame is inertial, and (as usual) gravity points down, opposite to $\underline{3^E}$.

The sensors output the below values:

$$\text{accelerometer: } \left[\underline{\alpha}\right]^B = \left\|\underline{g}\right\| \begin{bmatrix} -\frac{1}{2} \\ 0 \\ \frac{\sqrt{3}}{2} \end{bmatrix} \text{m/s}^2$$

$$\text{rate gyroscope: } \left[\underline{\gamma}\right]^B = \begin{bmatrix} 2 \\ 3 \\ 0 \end{bmatrix} \text{rad/s}$$

$$\text{optical flow sensor: } \begin{bmatrix} \sigma_1 \\ \sigma_2 \end{bmatrix} = \begin{bmatrix} -1 \\ 4 \end{bmatrix} \text{ rad/s}$$

$$\text{range sensor: } \delta = 2 \text{ m}$$

Assume that the sensors are ideal (that is, without noise, bias, scale or other errors). Given is that the yaw angle is zero, that is $\psi = 0$.

Find

(a) the pitch and roll angles θ, ϕ,
(b) the angular velocity $\left[\underline{\omega^{BE}}\right]^B$,
(c) the height at which the object is moving s_3, and
(d) the translational velocity of the object $\left[\underline{v_B^E}\right]^E$.

Hint: For the velocity, recall that the object is travelling horizontally and has zero velocity in the vertical direction.

Problem 5.9 (*7 points*) Your goal is to estimate the orientation of a UAV, constrained on a single-degree-of-freedom test rig, as shown in the photo below.

The vehicle's only degree of freedom is about the $\underline{1}^B$ axis (that is, the vehicle's roll angle). Your goal is to use data from the accelerometer and rate gyroscope to estimate the vehicle's orientation. You may assume that the beam holding the vehicle is perfectly horizontal, aligns with the $\underline{1}^B$ axis, and passes through the vehicle's center of mass.

Using C++ pseudo-code, write an estimator function for the vehicle roll angle. Your function must return a single `float` variable (corresponding to the current estimate of the angle), and must accept as argument six `float` variables, corresponding to three components of the accelerometer measurement, and three components of the rate gyroscope output.

The function is called at a rate 100 Hz with updated sensor measurements. Your pseudo-code must be clearly commented to explain your reasoning (you may write comments with math symbols). You may assume that the sensors are bias-free, and you may assume that the roll angle remains small. Make sure to explain any steps you take to reduce the effects of sensor noise.

References

Geneva P, Eckenhoff K, Lee W, Yang Y, Huang G, OpenVINS: a research platform for visual-inertial estimation. In: 2020 IEEE international conference on robotics and automation (ICRA). IEEE, pp 4666–4672

Gowda M, Manweiler J, Dhekne A, Choudhury RR, Weisz JD (2016) Tracking drone orientation with multiple GPS receivers. In: Proceedings of the 22nd annual international conference on mobile computing and networking, pp 280–293

Mueller MW, Hehn M, D'Andrea R (2017) Covariance correction step for Kalman filtering with an attitude. J Guid Control Dyn 40(9):2301–2306

Siegwart R, Nourbakhsh IR, Scaramuzza D (2011) Introduction to autonomous mobile robots. MIT Press

Chapter 6
Quadcopter Dynamics

Fource to be reckoned with.

We are now ready to describe the dynamics of a quadcopter, which we can then use to control such a vehicle. This will combine our aerodynamic knowledge from Chap. 1 with the modelling tools of Chaps. 2 through 6. Our quadcopter will be as shown below.

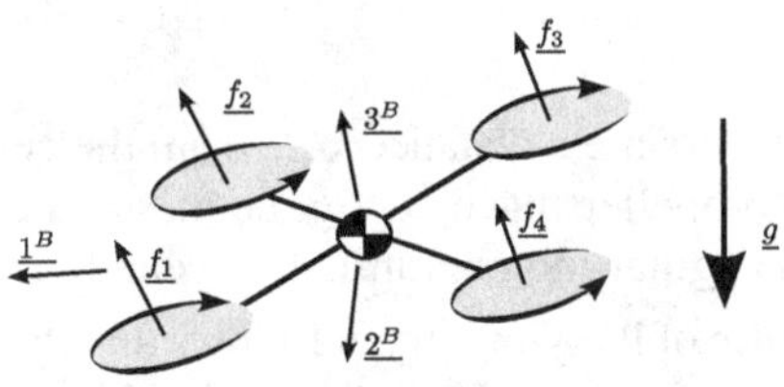

We define a body frame B, where $\underline{3}^B$ is perpendicular to the plane of the propellers, and $\underline{1}^B$ bisects two arms. The quadcopter is assumed to be rotationally symmetric, so that the arms are perpendicular to one another, and the propellers are all equally far from the vehicle's geometric center (which is also taken to coincide with its center of mass) B. We define our usual earth-fixed frame E, which we take to be inertial, and wherein $\underline{1}^E$ is a horizontal direction, and $\underline{3}^E$ points opposite to gravity (upwards).

The propeller P_i is displaced from the vehicle's center of mass by $\underline{s}_{P_i B}$, and produces a force $\underline{f}_{P_i}$, and a moment about its rotation axis $\underline{n}_{P_i}$. This convention is shown in Fig. 6.1.

M. W. Mueller, *Dynamics and Control of Autonomous Flight*, Mathematical Engineering,
https://doi.org/10.1007/978-3-031-91871-1_6

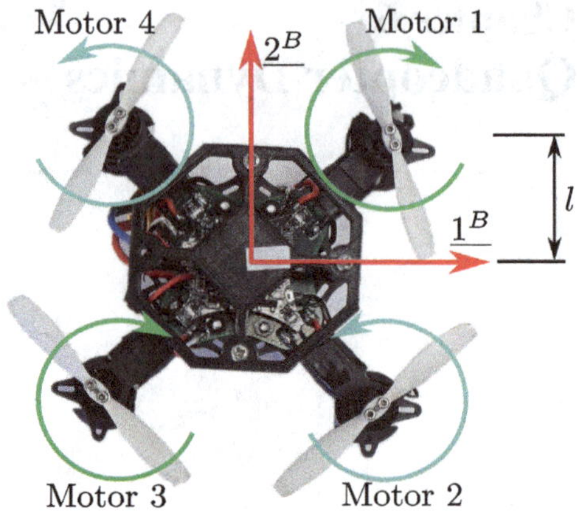

Fig. 6.1 The convention that we will use for the propeller numbering, frame, and size

The displacements are defined as below, introducing a scalar length l,

$$\underline{s_{P_1B}} = l\,(+\underline{1}^B + \underline{2}^B) \tag{6.1}$$

$$\underline{s_{P_2B}} = l\,(+\underline{1}^B - \underline{2}^B) \tag{6.2}$$

$$\underline{s_{P_3B}} = l\,(-\underline{1}^B - \underline{2}^B) \tag{6.3}$$

$$\underline{s_{P_4B}} = l\,(-\underline{1}^B + \underline{2}^B) \tag{6.4}$$

Note that thus the propellers are a distance $\sqrt{2}l$ from the center of mass.

We assume that the propellers have negligible mass (and thus neglect their mass moment of inertia and angular momentum). The vehicle has an inertia tensor $\underline{J_B^B}$, and has mass m^B. Because of its symmetry, it follows that the inertia, when expressed in the body-fixed B coordinate system, is diagonal, with only two unique entries:

$$\left[\underline{J_B^B}\right]^B = \begin{bmatrix} J_{xx} & 0 & 0 \\ 0 & J_{xx} & 0 \\ 0 & 0 & J_{zz} \end{bmatrix} \tag{6.5}$$

The orientation of the quadcopter with respect to the earth is given by the rotation tensor $\underline{R^{BE}}$, expressed using the transformation matrix $[T]^{EB}$. We fix a point on the Earth, for example where the vehicle took off, and call it T, so that the vehicle's displacement from that point is given by $\underline{s_{BT}}$.

6.1 Propeller Forces and Torques

Using the propeller thrust and torque model from Sect. 1.1.3, the force of each propeller is parallel to the body-fixed $\underline{3}^B$ axis: $\underline{f_{P_i}} = c_{P_i}\,\underline{3}^B$ where c_{P_i} is a scalar. The torque around the thrust axis is assumed to be proportional to the thrust, with the sign determined by the propeller's handedness (and conveniently expressed as $(-1)^{i+1}$), and having proportionality constant κ:

$$\underline{n_{P_i}} = (-1)^{i+1}\kappa\, \underline{f_{P_i}} = (-1)^{i+1}\kappa c_{P_i}\, \underline{3^B} \quad \text{for} \quad i \in \{1, 2, 3, 4\} \tag{6.6}$$

Finally, the vehicle is acted upon by gravity $\underline{g}$, of magnitude $\left\| \underline{g} \right\| \approx 9.81\,\mathrm{m\,s^{-2}}$. Our choice of E frame means that $\underline{g} = -\left\| \underline{g} \right\| \underline{3^E}$, or $\left[\underline{g} \right]^E \approx (0, 0, -9.81)\ \mathrm{m\,s^{-2}}$.

Now we can write down Newton's law, and use the simplification that the vehicle's mass is constant, and choosing the B point as the vehicle's center of mass:

$$D^E\, \underline{p_B^E} = m^B D^E\, \underline{v_B^E} = \underline{f} = \sum_{i=1}^{4} \underline{f_{P_i}} + m^B\, \underline{g} \tag{6.7a}$$

$$\Leftrightarrow m^B\, \underline{a_B^E} = \sum_{i=1}^{4} \underline{f_{P_i}} + m^B\, \underline{g} \tag{6.7b}$$

$$= \left(\sum_{i=1}^{4} c_{P_i} \right) \underline{3^B} + m^B\, \underline{g} \tag{6.7c}$$

Similarly, we can write down Euler's law, where we directly use the form for easy coordination in the body-fixed frame

$$\underline{J_B^B} D^B\, \underline{\omega^{BE}} + \underline{\Omega^{BE}}\, \underline{J_B^B}\, \underline{\omega^{BE}} = \underline{n_B} \tag{6.8a}$$

$$\underline{n_B} = \sum_{i=1}^{4} \left(\underline{K}\!\left(\underline{s_{P_i B}} \right) c_{P_i}\, \underline{3^B} + (-1)^{i+1}\kappa c_{P_i}\, \underline{3^B} \right) \tag{6.8b}$$

$$= \sum_{i=1}^{4} \left(c_{P_i} \left(\underline{K}\!\left(\underline{s_{P_i B}} \right) \underline{3^B} + (-1)^{i+1}\kappa\, \underline{3^B} \right) \right) \tag{6.8c}$$

where we've noted that the force acting at a distance also produces a moment, see (4.8). Substituting for $\underline{s_{P_i B}} = \pm \underline{1^B} \pm \underline{2^B}$, and expanding the vector product, we get

$$\begin{aligned} \underline{n_B} = {} & c_{P_1} \left(+l\, \underline{1^B} - l\, \underline{2^B} + \kappa\, \underline{3^B} \right) \\ & + c_{P_2} \left(-l\, \underline{1^B} - l\, \underline{2^B} - \kappa\, \underline{3^B} \right) \\ & + c_{P_3} \left(-l\, \underline{1^B} + l\, \underline{2^B} + \kappa\, \underline{3^B} \right) \\ & + c_{P_4} \left(+l\, \underline{1^B} + l\, \underline{2^B} - \kappa\, \underline{3^B} \right) \end{aligned} \tag{6.9}$$

We now coordinate in B, and will look at it term-by-term. On the left-hand side of (6.8a) we have

$$\left[\underline{J_B^B}\right]^B \left[D^B \underline{\omega^{BE}}\right] B = \begin{bmatrix} J_{xx}\dot{p} \\ J_{xx}\dot{q} \\ J_{zz}\dot{r} \end{bmatrix} \tag{6.10}$$

$$\left[\underline{\Omega^{BE}}\right]^B \left[\underline{J_B^B}\right]^B \left[\underline{\omega^{BE}}\right]^B = \begin{bmatrix} 0 & -r & q \\ r & 0 & -p \\ -q & p & 0 \end{bmatrix} \begin{bmatrix} J_{xx}p \\ J_{xx}q \\ J_{zz}r \end{bmatrix} = \begin{bmatrix} (J_{zz} - J_{xx})\, qr \\ (J_{xx} - J_{zz})\, pr \\ 0 \end{bmatrix} \tag{6.11}$$

The right-hand side of (6.8a), coordinated in B, solves to the below, where we've introduced the three scalars for convenience $\left[\underline{n_B}\right]^B =: (n_1, n_2, n_2)$.

$$\left[\underline{n_B}\right]^B = \begin{bmatrix} n_1 \\ n_2 \\ n_3 \end{bmatrix} = c_{P_1} \begin{bmatrix} l \\ -l \\ \kappa \end{bmatrix} + c_{P_2} \begin{bmatrix} -l \\ -l \\ -\kappa \end{bmatrix} + c_{P_3} \begin{bmatrix} -l \\ l \\ \kappa \end{bmatrix} + c_{P_4} \begin{bmatrix} l \\ l \\ -\kappa \end{bmatrix} \tag{6.12a}$$

$$= \begin{bmatrix} l & -l & -l & l \\ -l & -l & l & l \\ \kappa & -\kappa & \kappa & -\kappa \end{bmatrix} \begin{bmatrix} c_{P_1} \\ c_{P_2} \\ c_{P_3} \\ c_{P_4} \end{bmatrix} \tag{6.12b}$$

We note that we can achieve any moment by a suitable choice of forces c_{P_i}.

Combining (6.10)–(6.12b) and simplifying gives

$$\begin{bmatrix} l & -l & -l & l \\ -l & -l & l & l \\ \kappa & -\kappa & \kappa & -\kappa \end{bmatrix} \begin{bmatrix} c_{P_1} \\ c_{P_2} \\ c_{P_3} \\ c_{P_4} \end{bmatrix} = \begin{bmatrix} J_{xx}\dot{p} \\ J_{xx}\dot{q} \\ J_{zz}\dot{r} \end{bmatrix} + \begin{bmatrix} (J_{zz} - J_{xx})\, qr \\ (J_{xx} - J_{zz})\, pr \\ 0 \end{bmatrix} \tag{6.13}$$

We note that this is a coupled, nonlinear, differential equation for p, q, and r.

Next, we look at the translational dynamics, coordinating (6.7c) in the Earth-fixed frame:

$$m^B \left[\underline{a_B^E}\right]^E = \left(\sum_{i=1}^{4} c_{P_i}\right) \left[\underline{3^B}\right]^E + m^B \left[\underline{g}\right]^E \tag{6.14a}$$

$$= \left(\sum_{i=1}^{4} c_{P_i}\right) [T]^{EB} \left[\underline{3^B}\right]^B + m^B \left[\underline{g}\right]^E \tag{6.14b}$$

Defining $c_\Sigma := \sum_{i=1}^{4} c_{P_i}$; and $\left[\underline{a}_B^E\right]^E =: (a_1, a_2, a_3)$, we have

$$m^B \begin{bmatrix} a_1 \\ a_2 \\ a_3 \end{bmatrix} = c_\Sigma \, [T]^{EB} \begin{bmatrix} 0 \\ 0 \\ 1 \end{bmatrix} + m^B \begin{bmatrix} 0 \\ 0 \\ -\left\|\underline{g}\right\| \end{bmatrix} \tag{6.15}$$

We now have a compact description of how the dynamics of the drone are affected by the four motor forces: specifically, translation is dominated by the orientation $[T]^{EB}$ and the total thrust force, c_Σ. The orientation, in turn, is affected by the three-dimensional moment, which is made up of the four motor forces.

6.2 Transformation of the Inputs

We notice that the inputs (the quantities we can choose instantaneously) are c_{P_i}, but enter the dynamics in a very specific way: the translational acceleration is a function *only* of their sum, while the angular dynamics map the four-dimensional inputs linearly to three-dimensional torque vector. For this reason it is usually more convenient to use the torques $\left[\underline{n}_B\right]^B$ from (6.12b) and the total thrust c_Σ from (6.15) as inputs. The transformation between the two forms is linear:

$$\begin{bmatrix} c_\Sigma \\ n_1 \\ n_2 \\ n_3 \end{bmatrix} = \begin{bmatrix} 1 & 1 & 1 & 1 \\ l & -l & -l & l \\ -l & -l & l & l \\ \kappa & -\kappa & \kappa & -\kappa \end{bmatrix} \begin{bmatrix} c_{P_1} \\ c_{P_2} \\ c_{P_3} \\ c_{P_4} \end{bmatrix} \tag{6.16}$$

We can solve the forces as

$$\begin{bmatrix} c_{P_1} \\ c_{P_2} \\ c_{P_3} \\ c_{P_4} \end{bmatrix} = \frac{1}{4} \begin{bmatrix} 1 & l^{-1} & -l^{-1} & \kappa^{-1} \\ 1 & -l^{-1} & -l^{-1} & -\kappa^{-1} \\ 1 & -l^{-1} & l^{-1} & \kappa^{-1} \\ 1 & l^{-1} & l^{-1} & -\kappa^{-1} \end{bmatrix} \begin{bmatrix} c_\Sigma \\ n_1 \\ n_2 \\ n_3 \end{bmatrix}. \tag{6.17}$$

Often, the logical component that implements (6.17) is called the *mixer* or *mixer matrix*, which “mixes” the desired force and torque to compute the individual motor forces.

6.3 State-Space Description

We are now in a position to write the quadcopter's dynamics as a set of first-order differential equations, for control. We will define the state to consist of:

position of the quadcopter's center of mass as expressed in the earth-fixed coordinate system, $\left[\underline{s_{BT}}\right]^E =: (s_1, s_2, s_3)$ (recall that T was a fixed "target" point, fixed with respect to the earth);

velocity of the quadcopter's center of mass relative to the earth-fixed frame, expressed in the earth-fixed coordinate system, $\left[\underline{v_B^E}\right]^E =: (v_1, v_2, v_3)$;

orientation of the body-fixed frame with respect to the earth-fixed frame, as encoded in the transformation matrix

$$[T]^{EB} =: \begin{bmatrix} t_{11} & t_{12} & t_{13} \\ t_{21} & t_{22} & t_{23} \\ t_{31} & t_{32} & t_{33} \end{bmatrix} \tag{6.18}$$

angular velocity of the body-fixed frame with respect to the earth-fixed frame, expressed in the body-fixed coordinate system, $\left[\underline{\omega^{BE}}\right]^B = (p, q, r)$.

We will use the simplified input of the torque vector and the total force, noting that this can be converted back to motor forces using (6.17).

$$\frac{\mathrm{d}}{\mathrm{d}t}\begin{bmatrix} s_1 \\ s_2 \\ s_3 \end{bmatrix} = \begin{bmatrix} v_1 \\ v_2 \\ v_3 \end{bmatrix} \tag{6.19}$$

$$\frac{\mathrm{d}}{\mathrm{d}t}\begin{bmatrix} v_1 \\ v_2 \\ v_3 \end{bmatrix} = \frac{c_\Sigma}{m^B}\begin{bmatrix} t_{11} & t_{12} & t_{13} \\ t_{21} & t_{22} & t_{23} \\ t_{31} & t_{32} & t_{33} \end{bmatrix}\begin{bmatrix} 0 \\ 0 \\ 1 \end{bmatrix} + \begin{bmatrix} 0 \\ 0 \\ -\left\|\underline{g}\right\| \end{bmatrix} \tag{6.20}$$

$$\frac{\mathrm{d}}{\mathrm{d}t}\begin{bmatrix} t_{11} & t_{12} & t_{13} \\ t_{21} & t_{22} & t_{23} \\ t_{31} & t_{32} & t_{33} \end{bmatrix} = \begin{bmatrix} t_{11} & t_{12} & t_{13} \\ t_{21} & t_{22} & t_{23} \\ t_{31} & t_{32} & t_{33} \end{bmatrix}\begin{bmatrix} 0 & -r & q \\ r & 0 & -p \\ -q & p & 0 \end{bmatrix} \tag{6.21}$$

$$\frac{\mathrm{d}}{\mathrm{d}t}\begin{bmatrix} p \\ q \\ r \end{bmatrix} = \begin{bmatrix} n_1/J_{xx} \\ n_2/J_{xx} \\ n_3/J_{zz} \end{bmatrix} + \frac{J_{xx} - J_{zz}}{J_{xx}}\begin{bmatrix} qr \\ -pr \\ 0 \end{bmatrix} \tag{6.22}$$

Given a set of initial conditions, and inputs as a function of time, we can use (6.19)–(6.22) to solve for the full state trajectory of the quadcopter.[1]

[1] Note that you would typically not implement the matrix differential equation of (6.21), as it is much easier to employ the Euler Symmetric Parameters (AKA unit quaternion), which are discussed in Sect. 8.1.2. For analysis, though, we prefer the transformation matrix form.

6.3.1 Equilibrium and Linearization About Hover

For small deviations from hover, where the velocity, the orientation, and the angular velocity are all small, we can compute a simpler set of differential equations. We will first compute the equilibrium state of the quadcopter, that is, a state where it is at rest. The formal method for doing this is to set all the derivatives in (6.19)–(6.22) to zero. We denote by an asterisk the value of the states at equilibrium, and solve.

From (6.19) follows that $v_1^* = v_2^* = v_3^* = 0$, from (6.21) that $p^* = q^* = r^* = 0$. Substituting the rates into (6.22) we can conclude that the input torque must be zero, $m_1^* = m_2^* = m_3^* = 0$. We note that the right-hand side column of the transformation matrix is specified by (6.20), so that $t_{13} = t_{23} = 0$ and (because the columns are unit vectors) $t_{33} = 1$. Since the rows are also unit vectors, we have $t_{31} = t_{32} = 0$. Finally, we can conclude that the total thrust must be the weight of the vehicle, $c_\Sigma^* = m^B \left\| \underline{g} \right\|$.

For the linearization, it will be more convenient to express the transformation matrix using the Euler yaw-pitch-roll (ψ-θ-ϕ) sequence. The equilibrium values can be computed by substituting our knowledge for t_{ij}^*:

$$\begin{bmatrix} t_{11}^* & t_{12}^* & 0 \\ t_{21}^* & t_{22}^* & 0 \\ 0 & 0 & 1 \end{bmatrix} = \begin{bmatrix} \mathrm{c}\psi^*\mathrm{c}\theta^* & \mathrm{c}\psi^*\mathrm{s}\theta^*\mathrm{s}\phi^* - \mathrm{s}\psi^*\mathrm{c}\phi^* & \mathrm{c}\psi^*\mathrm{s}\theta^*\mathrm{c}\phi^* - \mathrm{s}\psi^*\mathrm{s}\phi^* \\ \mathrm{s}\psi^*\mathrm{c}\theta^* & \mathrm{s}\psi^*\mathrm{s}\theta^*\mathrm{s}\phi^* + \mathrm{c}\psi^*\mathrm{c}\phi^* & \mathrm{s}\psi^*\mathrm{s}\theta^*\mathrm{c}\phi^* - \mathrm{c}\psi^*\mathrm{s}\phi^* \\ -\mathrm{s}\theta^* & \mathrm{c}\theta^*\mathrm{s}\phi^* & \mathrm{c}\theta^*\mathrm{c}\phi^* \end{bmatrix} \tag{6.23}$$

From this we can immediately conclude that $\theta^* = 0$ and $\phi^* = 0$ (which also matches our intuition).

We note that the equilibrium does not specify a position s_i^*, or the yaw angle ψ^*. Intuitively, this makes sense: the quadcopter can hover anywhere in the world, and at any yaw angle. For simplicity, we will now make the additional restriction that $\psi^* = 0$, i.e. we'll look at hovering with small yaw angles. Note that this is without loss of generality—if you wanted to hover at a different angle, simply create a new world-fixed coordinate system wherein that yaw angle is zero.

We now linearize the dynamics (6.19)–(6.22) at these values, and use the Euler angle dynamics for (6.21). We introduce the input deviation Δc_Σ, so that $c_\Sigma =: c_\Sigma^* + \Delta c_\Sigma$ (since all other quantities are zero at the equilibrium, we don't need to explicitly identify deviations).

For a small angles we make the substitution that $\sin\theta \approx \theta$, $\cos\theta \approx 1$, and that a product of any two deviations is zero, e.g. $\phi\theta \approx 0$. We can now compute the small-angles transformation matrix:

$$[T]^{EB} = \begin{bmatrix} \mathrm{c}\psi\mathrm{c}\theta & \mathrm{c}\psi\mathrm{s}\theta\mathrm{s}\phi - \mathrm{s}\psi\mathrm{c}\phi & \mathrm{c}\psi\mathrm{s}\theta\mathrm{c}\phi - \mathrm{s}\psi\mathrm{s}\phi \\ \mathrm{s}\psi\mathrm{c}\theta & \mathrm{s}\psi\mathrm{s}\theta\mathrm{s}\phi + \mathrm{c}\psi\mathrm{c}\phi & \mathrm{s}\psi\mathrm{s}\theta\mathrm{c}\phi - \mathrm{c}\psi\mathrm{s}\phi \\ -\mathrm{s}\theta & \mathrm{c}\theta\mathrm{s}\phi & \mathrm{c}\theta\mathrm{c}\phi \end{bmatrix} \tag{6.24a}$$

$$\approx \begin{bmatrix} 1 & -\psi & \theta \\ \psi & 1 & -\phi \\ -\theta & \phi & 1 \end{bmatrix} \tag{6.24b}$$

For the derivative of the velocity, Eq. (6.20), we note that we need the product $[T]^{EB}\left[\underline{3}^{B}\right]^{B}$:

$$\begin{bmatrix} t_{11} & t_{12} & t_{13} \\ t_{21} & t_{22} & t_{23} \\ t_{31} & t_{32} & t_{33} \end{bmatrix} \begin{bmatrix} 0 \\ 0 \\ 1 \end{bmatrix} \approx \begin{bmatrix} 1 & -\psi & \theta \\ \psi & 1 & -\phi \\ -\theta & \phi & 1 \end{bmatrix} \begin{bmatrix} 0 \\ 0 \\ 1 \end{bmatrix} = \begin{bmatrix} \theta \\ -\phi \\ 1 \end{bmatrix} \tag{6.25}$$

and we substitute $c_\Sigma = m^B \left\| \underline{g} \right\| + \Delta c_\Sigma$

$$\frac{\mathrm{d}}{\mathrm{d}t} \begin{bmatrix} v_1 \\ v_2 \\ v_3 \end{bmatrix} \approx \frac{m^B \left\| \underline{g} \right\| + \Delta c_\Sigma}{m^B} \begin{bmatrix} \theta \\ -\phi \\ 1 \end{bmatrix} + \begin{bmatrix} 0 \\ 0 \\ -\left\| \underline{g} \right\| \end{bmatrix} \approx \begin{bmatrix} \theta \left\| \underline{g} \right\| \\ -\phi \left\| \underline{g} \right\| \\ \frac{\Delta c_\Sigma}{m^B} \end{bmatrix} \tag{6.26}$$

We can make the same substitutions and simplifications for the kinematics of the Euler angles:

$$\frac{\mathrm{d}}{\mathrm{d}t} \begin{bmatrix} \phi \\ \theta \\ \psi \end{bmatrix} = \begin{bmatrix} 1 & \sin\phi \tan\theta & \cos\phi \tan\theta \\ 0 & \cos\phi & -\sin\phi \\ 0 & \sin\phi / \cos\theta & \cos\phi / \cos\theta \end{bmatrix} \begin{bmatrix} p \\ q \\ r \end{bmatrix} \tag{6.27a}$$

$$\approx \begin{bmatrix} 1 & 0 & \theta \\ 0 & 1 & -\phi \\ 0 & \phi & 1 \end{bmatrix} \begin{bmatrix} p \\ q \\ r \end{bmatrix} \approx \begin{bmatrix} p \\ q \\ r \end{bmatrix} \tag{6.27b}$$

Since the angular velocity is small, we neglect the quadratic term in the angular velocity dynamics (the terms with pq), and we can make all the substitutions and write the dynamics of the quadcopter as:

$$\frac{\mathrm{d}}{\mathrm{d}t} \begin{bmatrix} s_1 \\ s_2 \\ s_3 \end{bmatrix} = \begin{bmatrix} v_1 \\ v_2 \\ v_3 \end{bmatrix} \tag{6.28}$$

$$\frac{\mathrm{d}}{\mathrm{d}t} \begin{bmatrix} v_1 \\ v_2 \\ v_3 \end{bmatrix} \approx \begin{bmatrix} \theta \left\| \underline{g} \right\| \\ -\phi \left\| \underline{g} \right\| \\ \frac{\Delta c_\Sigma}{m^B} \end{bmatrix} \tag{6.29}$$

$$\frac{\mathrm{d}}{\mathrm{d}t} \begin{bmatrix} \phi \\ \theta \\ \psi \end{bmatrix} \approx \begin{bmatrix} p \\ q \\ r \end{bmatrix} \tag{6.30}$$

$$\frac{\mathrm{d}}{\mathrm{d}t} \begin{bmatrix} p \\ q \\ r \end{bmatrix} \approx \begin{bmatrix} n_1 / J_{xx} \\ n_2 / J_{xx} \\ n_3 / J_{zz} \end{bmatrix} \tag{6.31}$$

This is extremely convenient, and we notice that it decouples into four simpler systems: one for each horizontal direction, one for the height, and one for the yaw angle. Each system has an own scalar input, furthermore.

$$\frac{\mathrm{d}}{\mathrm{d}t}\begin{bmatrix} s_1 \\ v_1 \\ \theta \\ q \end{bmatrix} = \begin{bmatrix} v_1 \\ \theta \left\| \underline{g} \right\| \\ q \\ n_2/J_{xx} \end{bmatrix} \tag{6.32}$$

$$\frac{\mathrm{d}}{\mathrm{d}t}\begin{bmatrix} s_2 \\ v_2 \\ \phi \\ p \end{bmatrix} = \begin{bmatrix} v_2 \\ -\phi \left\| \underline{g} \right\| \\ p \\ n_1/J_{xx} \end{bmatrix} \tag{6.33}$$

$$\frac{\mathrm{d}}{\mathrm{d}t}\begin{bmatrix} s_3 \\ v_3 \end{bmatrix} = \begin{bmatrix} v_3 \\ \Delta c_\Sigma/m^B \end{bmatrix} \tag{6.34}$$

$$\frac{\mathrm{d}}{\mathrm{d}t}\begin{bmatrix} \psi \\ r \end{bmatrix} = \begin{bmatrix} r \\ n_3/J_{zz} \end{bmatrix} \tag{6.35}$$

This description is, of course, only valid to first order, but is illuminating. The complex dynamics of a quadcopter can be broken into four separate sub-systems. Two of these (the horizontal position components) are (up to scaling) each an input integrated four times. The remaining two sub-systems (height and yaw angle) are each an input integrated twice (up to a scaling). We may thus think of controlling a quadcopter as controlling these four separate sub-systems, each of which is simple to describe. The topic of how to choose control inputs is discussed in Chap. 7.

Problems for Chapter 6

Problem 6.1 (*10 points*) Consider the following quadcopter design, where c_1 through c_4 are the motor forces that we can select for the propellers as indicated. All distance are from the center of mass, and the vehicle is mirror-symmetric about the $\underline{1}^B$ axis. The propeller torque constant (mapping from thrust value to moment about the rotation axis) is $\kappa = 0.016\,\mathrm{m}$.

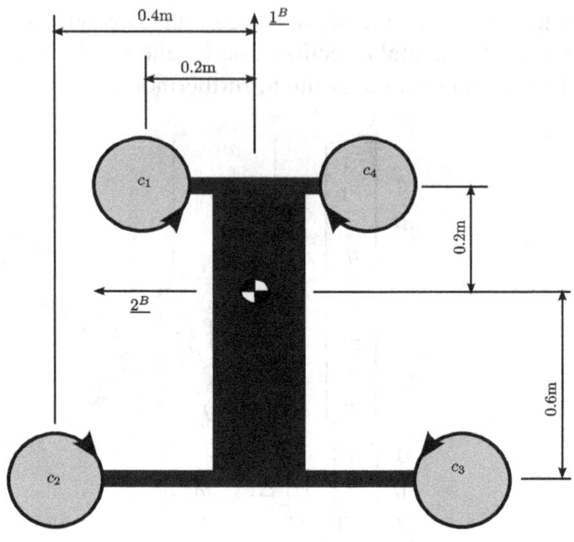

Compute the mixer matrix M so that

$$\begin{bmatrix} c_1 \\ c_2 \\ c_3 \\ c_4 \end{bmatrix} = M \begin{bmatrix} c_\Sigma \\ n_1 \\ n_2 \\ n_3 \end{bmatrix} \tag{6.36}$$

where c_Σ is a given total thrust command, and $\left[\underline{n_B}\right]^B = \left[n_1, n_2, n_3\right]$ is a given moment.

Note: Feel free to use a computer to invert matrices, as needed.

Problem 6.2 (*12 points*) Consider a quadcopter with the following physical parameters:

$$\begin{aligned} \text{Scale:}\quad & l = 33 \times 10^{-3}\,\text{m} \\ \text{Mass:}\quad & m^B = 32 \times 10^{-3}\,\text{kg} \\ \text{Inertia tensor:}\quad & \left[\underline{J_B^B}\right]^B = \begin{bmatrix} 16 & 0 & 0 \\ 0 & 16 & 0 \\ 0 & 0 & 29 \end{bmatrix} \times 10^{-6}\,\text{kgm}^2 \\ \text{Propeller torque constant:}\quad & \kappa = 0.01\,\text{m} \end{aligned}$$

The vehicle is initially at hover, and you are given a desired angular acceleration:

$$\left[\frac{\mathrm{d}}{\mathrm{d}t}\underline{\omega}\right]^B = \begin{bmatrix} 30 \\ 0 \\ 10 \end{bmatrix} \text{rad}\,\text{s}^{-2} \tag{6.37}$$

which is to be achieved without changing the vertical velocity. Compute the required motor forces c_1, c_2, c_3, and c_4.

Problem 6.3 (*10 points*) For the quadcopter with parameters as in Problem 6.2, assume that the motors can produce a force in the range $c_i \in [0, 0.13]$N. What is the maximum angular acceleration about the body-fixed $\underline{3}^B$ axis that the vehicle can achieve from hover, at constant height?

Problem 6.4 (*8 points*) A colleague has proposed the following modification to a quadcopter, where the sense of rotation of motors 2 and 3 have been swapped from the default (as well as their propellers):

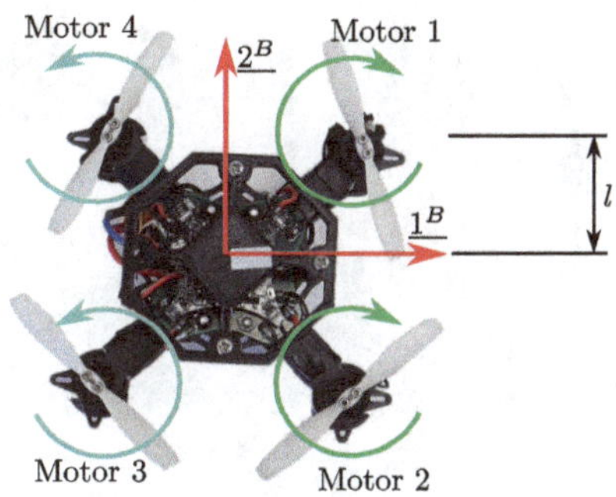

Using the mixer matrix M, argue why this configuration is problematic, compared to the standard configuration as presented in Sect. 6.2.

Problem 6.5 (*8 points*) A hovering quadcopter is initially stationary. Motor number 3 fails, and suddenly produces zero force/torque. The vehicle parameters are as given below, where we use the same motor numbering as in Fig. 6.1.

$$m^B = 30 \times 10^{-3}\,\text{kg}$$
$$l = 30 \times 10^{-3}\,\text{m}$$
$$\kappa = 0.01\,\text{m}$$
$$\left[\underline{I_B^B}\right]^B = \begin{bmatrix} 20 & 0 & 0 \\ 0 & 20 & 0 \\ 0 & 0 & 30 \end{bmatrix} \times 10^{-6}\,\text{kgm}^2$$

(a) Find the instantaneous acceleration of the quadcopter $\left[\underline{a_B^E}\right]^E$.

(b) Find the instantaneous angular acceleration of the quadcopter $\left[\frac{\mathrm{d}}{\mathrm{d}t}\underline{\omega^{BE}}\right]^B$.

Chapter 7
Control and Stabilization

Hard to fly safely if you are not stable.

At this point, we have a system of which we understand the dynamics, its inputs, and its outputs. By this, we mean that there are certain things that we get to choose ('inputs') that affect the system internally (through its 'dynamics') and cause the system to produce measurable signals ('outputs'). Note that, in the context of control systems, we use 'dynamics' in a broader sense than when we say dynamics in an engineering mechanics context as in Chap. 4. In controls, a dynamic system is any system that evolves over time, and the system's dynamics are the equations that govern this change.

7.1 Language, Symbols, and Equations

We will consider only continuous-time systems here, and will use the generic symbols that you're likely to find in most controls texts. A typical dynamic system may be written as follows:

$$\dot{x} = f(x, u) \tag{7.1}$$

wherein

- x is a function of time, specifically the system *state*, and is usually vector-valued (that is, it is a *collection of quantities*);
- $\dot{x} = \frac{\mathrm{d}}{\mathrm{d}t}x$ is the derivative of that state, the vector of element-wise derivatives of x;
- u is the input, also usually vector-valued; and

M. W. Mueller, *Dynamics and Control of Autonomous Flight*, Mathematical Engineering,
https://doi.org/10.1007/978-3-031-91871-1_7

- $f(\cdot, \cdot)$ is the system's governing equation, a function that maps from the state and the input to the derivative of the state.

We note that x and u are functions of time, and we typically use the short-hand x when we mean $x(t)$, i.e. a more verbose expression of (7.1) would be $\dot{x}(t) = f(x(t), u(t))$.[1]

The *state* of a system contains all of the quantities necessary so that we may predict what the system will do in the future. The *inputs* are typically things that we can choose, e.g. motor voltages, steering angles, etc. Given a physical system to model, the choice of state and input variables is somewhat of an art, and there are no correct answers. Different choices represent different trade-offs between tractability and error; this process is called "modelling" and we will not spend much time discussing it here.[2]

The purpose of *controls* is to manipulate the system (through the choice of inputs) so that we get some desirable behavior. For example, when controlling a car we aim to choose a steering wheel angle that maintains the car's position near the center of the lane we're driving in. For our generic system as given in (7.1) this can be expressed as choosing an input u so that the state x evolves in a desirable fashion. One of the most common objectives is to *regulate* the state so that it remains near an *equilibrium*, defined next.

7.1.1 Equilibria

A dynamic system ($\dot{x} = f(x, u)$) is said to have an equilibrium state and input pair x^* and u^* if

$$\dot{x} = f(x^*, u^*) = 0 \tag{7.2}$$

In other words, if the system state is at $x = x^*$, and we apply the input $u = u^*$, then the system state is constant (since $\dot{x} = 0$) and the system will remain at x^*.

We have seen this for the specific case of our quadcopter, where the vehicle is at equilibrium when it is hovering. We have already seen, for the quadcopter, that the equilibrium need not be unique (recall, the quadcopter can hover at any position in space, and at any yaw angle). In fact, in general, a system may have none, one, multiple (but finite), or an infinite number of equilibria.

Usually, there is one equilibrium of particular interest (i.e. that point where we want the system to remain). We usually redefine our state and inputs so that this point

---

[1] Sometimes the dynamics function f itself changes over time, and we'd have $\dot{x}(t) = f(t, x(t), u(t))$. See "Autonomous" versus "Non-autonomous" dynamic systems. For this course, we will stick to "autonomous" systems, where f has no explicit time dependence.

[2] Remember—all models are wrong, but some models are useful. For mechanical systems, like our UAV, the model is often relatively easy to choose (in our case, treat the quadcopter as a rigid body with six degrees of freedom).

is at the origin—let x^*, u^* be the equilibrium of interest. Then define $\tilde{x} := x - x^*$, and $\tilde{u} := u - u^*$. We then have:

$$\begin{aligned}\dot{\tilde{x}} &= \dot{x} - \dot{x}^* && (7.3a)\\ &= f(x,u) - 0 && (7.3b)\\ &= f(\tilde{x} + x^*, \tilde{u} + u^*) && (7.3c)\\ &=: \tilde{f}(\tilde{x}, \tilde{u}) && (7.3d)\end{aligned}$$

so that we can now work with the new system $\dot{\tilde{x}} = \tilde{f}(\tilde{x}, \tilde{u})$ for which $\tilde{f}(0, 0) = 0$.

7.1.2 Linear Versus Nonlinear Systems

A system is said to be *linear* if we can find matrices A and B so that

$$\dot{x} = f(x,u) = Ax + Bu \tag{7.4}$$

Linear systems are much easier to analyze, and linear systems theory provides us with incredibly powerful tools to analyze such systems.

A nonlinear system is any system which is not linear—this is a much broader class, and such systems can be much more complex than linear systems. We typically do not have general results for such systems.

Note that, in reality, nothing is linear—for any system, there will be some physical limit that prevents us from applying arbitrarily large inputs, such as power limitations. Nonetheless, "everything is linear to first order".[3] When a linear model is an acceptable approximation for our true physics, it would be foolish to not exploit that.

7.2 Control of Simple Systems

As we noted, the purpose of controls is to manipulate the system (by choosing inputs) so that we achieve some desired behavior. Most commonly, we choose our input u in response to the state x, through a *feedback law*, so that we design a function $k(\cdot)$, and implement $u = k(x)$ so that the *closed-loop system* below satisfies our design goals:

$$\dot{x} = f(x,u) = f(x, k(x)) =: \bar{f}(x) \tag{7.5}$$

In other words, we wish to manipulate the differential equation $f(\cdot,\cdot)$ into a new ODE ($\dot{x} = \bar{f}(x)$) which satisfies our design goal. Most of what we do in controls

[3] A somewhat tongue-in-cheek comment of Prof. Raffaello D'Andrea.

engineering is manipulating differential equations. We will see this in action for some simple examples below. In each case, we will start with a system $\dot{x} = f(x, u)$ and set ourselves the objective of designing a control law $u = k(x)$ so that the state x is driven to some desired state *setpoint* $x = x_s$.

We begin by looking at two simple types of systems: single and double integrators. It turns out that many real systems can be approximated well by such models; moreover their control design can be done in a very intuitive fashion.

7.2.1 Single Integrator

Consider a scalar system with state x and scalar input u, which evolves according to the differential equation

$$\dot{x} = f(x, u) = u \tag{7.6}$$

We note that the equilibrium is any state value x paired with $u = 0$. This is called a "single integrator" as the state integrates the input once.

In the simplest case, our goal might be to regulate this system so that the state converges to the equilibrium, that is, $x \to 0$. We will choose a *proportional controller*, so that the input is proportional to the state, and specifically we choose $u = k(x) = ax$, where a is a scalar constant. The resulting closed-loop ODE is then

$$\dot{x} = f(x, u) = f(x, ax) = ax \tag{7.7}$$

This is, fortunately, easy to solve for in closed form, and we can solve the state at any time t, $x(t)$, if we know the value it has at time $t = 0$, (the initial value) $x_0 = x(0)$:

$$x(t) = e^{at} x_0 \tag{7.8}$$

Since x_0 is fixed, we know that $x(t) \to 0$ only if $e^{at} \to 0$; this will happen only if a is negative. This type of control is more conveniently specified in terms of a *time constant* τ, so that $a = -\frac{1}{\tau}$, and we have

$$u = -\frac{1}{\tau}x, \qquad x(t) = e^{-t/\tau} x_0 \tag{7.9}$$

We note that this intuitive, because the time constant encodes how fast the system responds—at $t = \tau$, we have that $x(\tau) \approx 0.368x_0$, or about 36% of the initial error remains after the duration of one time constant τ. After 2τ, the error is reduced to about 14%, and after 3τ to 5%. A plot of the state and input are graphed against time in Fig. 7.1, where we see that a shorter time constant causes the system to converge more quickly, but requires larger peak inputs.

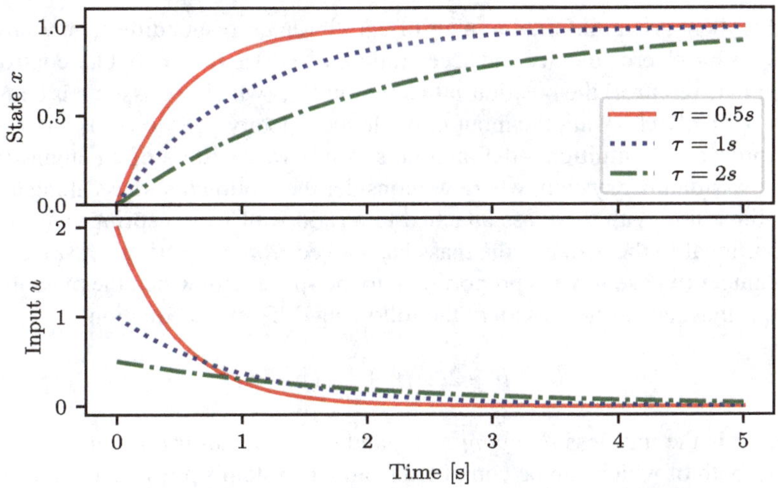

Fig. 7.1 Closed-loop response of a first-order system with varying control parameters

Of course, if we want the system to converge to an arbitrary setpoint x_s (which may be non-zero), we can apply the control

$$u = -\frac{1}{\tau}(x - x_s) \tag{7.10}$$

so that the input is proportional to the state error.

7.2.2 Double Integrator

Possibly the simplest system, after a single integrator, is the double integrator. Such systems are very common: after all, Newton's law (e.g.) describes position as the double integral of force (scaled through the mass).

Consider a two-state system with state containing position p and velocity v, and a scalar acceleration input u, which evolves according to the differential equation

$$\dot{p} = v \qquad \text{and} \qquad \dot{v} = u \tag{7.11}$$

One example of such a system is a train on a track, where the input u is the acceleration along the track.

First, note that we can easily convert this to the standard equations:

$$x := \begin{bmatrix} p \\ v \end{bmatrix}; \qquad \dot{x} = \begin{bmatrix} \dot{p} \\ \dot{v} \end{bmatrix} = \begin{bmatrix} v \\ u \end{bmatrix} =: f(x, u) \tag{7.12}$$

Note that there are an infinite number of equilibria, corresponding to any position, but always have zero velocity and acceleration: $v^* = 0$ and $u^* = 0$. Our control goal is to regulate (control) the position p to a setpoint p_s (which we assume is constant), by appropriately choosing the input u, while the velocity goes to zero.

We proceed by intuition—the favorite second-order system of an engineer[4] is a spring-mass-damper system, where we consider the motion of a mass along a single spatial direction, with the mass attached to a rigid wall with a spring (whose force is proportional to the distance the mass has moved from the spring's rest length p_s) and a damper (whose force is proportional to the speed with which the mass moves). We know that such systems satisfy the following differential equation

$$0 = \ddot{p} + 2\zeta\omega_n\dot{p} + \omega_n^2 (p - p_s) \tag{7.13}$$

wherein ζ is the unit-less *damping ratio*, and ω_n is the *natural frequency* (in units rad s^{-1}), both of which can be computed from the system's physical parameters.

Noting that we can substitute the input and speed into this ODE, we can solve the input u required so that the system (7.12) behaves like a spring-mass-damper system, with the damping ratio and natural frequency left to be defined by the controls engineer:

$$u = -2\zeta\omega_n v - \omega_n^2 (p - p_s) \tag{7.14}$$

This results in a proportional feedback law.

The natural frequency defines the "speed" of the system, while the damping ratio ζ defines the "shape" of the response. A typical choice of is $\zeta = \sqrt{1/2} \approx 0.7$, which minimizes the system's settling time for a given ω_n (the settling time is the time required for the system to be "near" the setpoint).

For a setpoint $x_s = 1$, and initial condition $x(0) = 0$, the closed-loop trajectories are shown for different values of natural frequency ω_n and damping ratio ζ in Fig. 7.2. We note that larger natural frequencies lead to "faster" and "more aggressive" behavior. The choice of damping ratio has little impact on the magnitude of the input, but we note that the position trajectory settles most quickly for $\zeta = 0.7$. As for the single integrator, there is a practical trade-off between rapid convergence, acceptable input values, and noise amplification.

This is called a *proportional-derivative (PD) controller* of the position, as the input has a part that is proportional to the position error, and one that's proportional to the velocity (derivative of position). Of course, the input u is still simply proportional to the state vector x.

[4] At least, the favorite of most mechanical engineers.

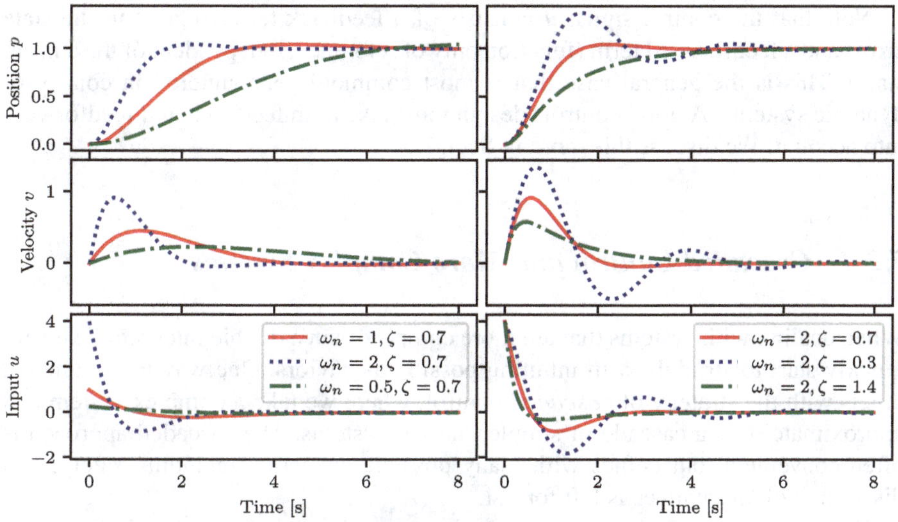

Fig. 7.2 Closed-loop response of a second-order system with varying control parameters

7.2.3 Feedforward Signals

We have now designed controllers which will bring the system to a stationary reference, for both single- and double-integrators. Often, though, the reference itself changes over time (think of a UAV filming a racecar, and the objective is to remain over the race car). In this case, we have $\dot{x}_s \neq 0$, and our objective becomes to have $x(t) \to x_s(t)$. To achieve this, it turns out we must modify our controller.

We define the *tracking error* as $e := x - x_s$—if $x(t) \to x_s(t)$, then clearly $e \to 0$, and it turns out to be more convenient to analyze $e(t)$ than $x(t)$.

Consider, for example a single integrator. If we were to apply (naïvely) our previous controller for a first order system, $u = -\frac{1}{\tau}(x - x_s) = -\frac{1}{\tau}e$, we have:

$$\dot{e} = \dot{x} - \dot{x}_s = u - \dot{x}_s \tag{7.15a}$$

$$= -\frac{1}{\tau}e - \dot{x}_s \tag{7.15b}$$

which will, in general, not converge to zero, because of the term $\dot{x}_s$. To correct for this, we must add a *feed-forward* term to our controller, and apply the input $u = -\frac{1}{\tau}(x - x_s) + \dot{x}_s$, analyzing now the error dynamics we have

$$\dot{e} = -\frac{1}{\tau}e \tag{7.16}$$

i.e. the error will decay to zero exponentially, with time constant τ.

Note that the control signal u consists of a feedback term (related to the state error) and a feedforward term (function only of reference, independent of the current state). This is the general case that is most commonly encountered in control of dynamic systems. A good controls design must take both feedback and feedforward into account. We discuss this more in Sect. 7.6.

7.2.4 Cascaded Control and More Complex Systems

When dealing with systems that are more complex than a double integrator, we very quickly start to struggle with intuiting possible behaviors. One way to circumvent this is with the strategy of *cascaded* control, where we take a complex system and approximate it as a cascade of simpler, nested systems. The cascaded approach is often convenient, but comes with many possible concerns (including safety!)—a discussion of alternatives is left for 7.4.

Consider, as a first example, using this approach to control a double integrator (rather than our intuition-based, mass-spring-damper control of Sect. 7.2.2), where we have

$$\dot{p} = v, \qquad \dot{v} = u. \tag{7.17}$$

and wish to control the position to p_{des}.

We start by assuming that, from "somewhere", we get a desired velocity v_{des} which will accomplish our control goal. Given this desired velocity, we know that a successful acceleration input would be

$$u = -\frac{1}{\tau_v}\left(v - v_{\text{des}}\right) \tag{7.18}$$

with the time constant τ_v a tuning constant. We now expect, if v_{des} is constant, that $v \to v_{\text{des}}$.

Zooming out, if this convergence ($v \to v_{\text{des}}$) happens sufficiently rapidly (i.e. the time constant τ_v is sufficiently fast (small)), we can treat the position as a single integrator with input v_{des}, which we can compute as a first-order feedback:

$$v_{\text{des}} = -\frac{1}{\tau_p}\left(p - p_{\text{des}}\right) \tag{7.19}$$

We can now keep doing this, as our systems become more complex. For a triple integrator, for example, where

$$\dot{x}_1 = x_2, \qquad \dot{x}_2 = x_3, \qquad \dot{x}_3 = u \tag{7.20}$$

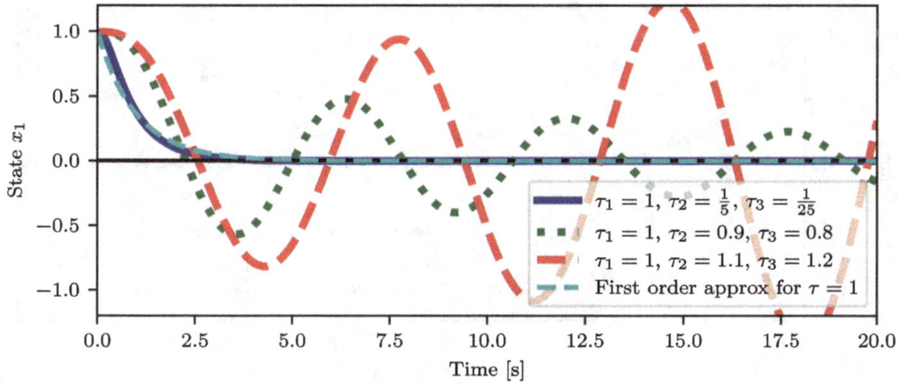

Fig. 7.3 Closed-loop response of a triple integrator with cascaded control approach

with goal $x_1 \to x_{1,\text{des}}$. This can be done with three cascaded single-integrator systems:

$$x_{2,\text{des}} = -\frac{1}{\tau_1}\left(x_1 - x_{1,\text{des}}\right) \tag{7.21}$$

$$x_{3,\text{des}} = -\frac{1}{\tau_2}\left(x_2 - x_{2,\text{des}}\right) \tag{7.22}$$

$$u = -\frac{1}{\tau_3}\left(x_3 - x_{3,\text{des}}\right) \tag{7.23}$$

Three implementations of this control strategy are shown in Fig. 7.3, as well as a first-order system with the same outermost time constant. As expected, when the time constants are suitably separated, i.e. $\tau_1 \gg \tau_2 \gg \tau_3$, then the third-order system behaves very similarly to the first-order. However, when the time constants are not well separated, performance is worse (large oscillations that decay only slowly), or even unstable if we have $\tau_1 < \tau_2 < \tau_3$.

If we go one step further, to a quadruple integrator, an interesting behavior appears. Such a system would be written as

$$\dot{x}_1 = x_2, \qquad \dot{x}_2 = x_3, \qquad \dot{x}_3 = x_4, \qquad \dot{x}_4 = u \tag{7.24}$$

with goal $x_1 \to x_{1,\text{des}}$. This can be done with three cascaded single-integrator systems:

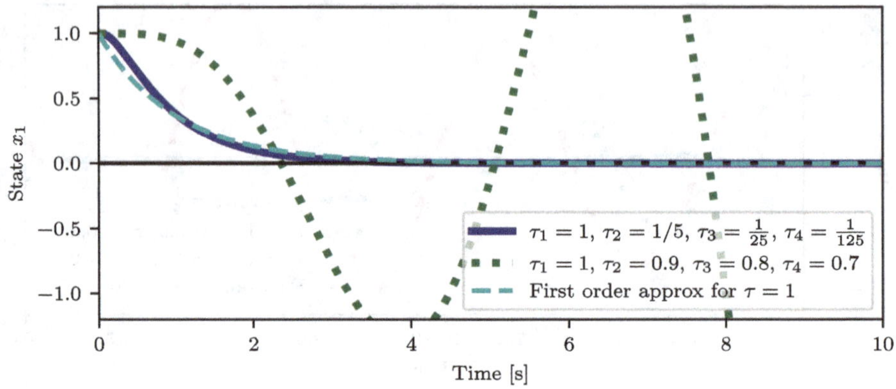

Fig. 7.4 Closed-loop response of a quadruple integrator with cascaded control approach

$$x_{2,\text{des}} = -\frac{1}{\tau_1}\left(x_1 - x_{1,\text{des}}\right) \tag{7.25}$$

$$x_{3,\text{des}} = -\frac{1}{\tau_2}\left(x_2 - x_{2,\text{des}}\right) \tag{7.26}$$

$$x_{4,\text{des}} = -\frac{1}{\tau_3}\left(x_3 - x_{3,\text{des}}\right) \tag{7.27}$$

$$u = -\frac{1}{\tau_4}\left(x_4 - x_{4,\text{des}}\right) \tag{7.28}$$

Some simulations are shown in Fig. 7.4, where we see again that if $\tau_4 \ll \tau_3 \ll \tau_2 \ll \tau_1$ we have that the system performs well and closely approximates a first-order system with time constant τ_1. However, we notice that the closed loop system can become unstable even if $\tau_4 < \tau_3 < \tau_2 < \tau_1$, i.e. if the time constants become too close to one another. This is, of course, concerning—when I'm designing a control system, I want to be very confident that the closed-loop system remains stable, and that the error smoothly goes to zero. If there's a danger of the system exploding, what are we to do?

7.2.5 *Avoiding Cascaded Control*

Based off the observation that a cascaded control can be unstable, if the separation between time constants is not large enough, one may be tempted to simply avoid cascaded control altogether (perhaps in favor of one of the strategies discussed in Sect. 7.4). However, this is not practically possible typically. Consider, for example, the height control of a quadcopter, taken from (6.34), repeated here:

$$\frac{\mathrm{d}}{\mathrm{d}t}\begin{bmatrix} s_3 \\ v_3 \end{bmatrix} = \begin{bmatrix} v_3 \\ \Delta c_\Sigma / m^B \end{bmatrix} \tag{7.29}$$

Setting $x_1 = s_3$, $x_2 = v_3$, and $u = \Delta c_\Sigma / m^B$, we have $\ddot{x}_1 = u$, i.e., a double integrator. We can then design a naïve double-integrator controller, as in Sect. 7.2.2, and conclude that the system will be stable for any choice of natural frequency and damping ratio.

However, for any real system, the total excess thrust Δc_Σ cannot be changed instantaneously. Recall from (1.14c) that each propeller's thrust is related to its angular velocity, $c_{P_i} \sim \Omega_i^2$. A change in thrust thus requires a change in angular velocity—however, the angular velocity of any real motors and propellers (with mass and thus inertia) can only change in a continuous fashion. One may perhaps model the motors as another single integrator, with the motor angular velocity approaching the desired angular velocity with a fast time constant, so that the height sub-system is in fact a triple integrator, where this third time constant is one that we cannot directly set. There may be more dynamics to consider—accelerating the motors requires torque, which is typically proportional to the current flowing through the motors. This current can also not change instantaneously, due to capacitance and inductance in the electric circuits. We thus see, very quickly, that almost any complex engineered system will have some cascaded control 'in the background'.

7.3 Stability

In the last section, we saw that closing the loop on the system ideally leads to the error converging to zero; however sometimes we saw that we can cause the system to "blow up", and errors to become arbitrarily large. We now develop the language to describe such phenomena precisely, using concepts of Lyapunov. We will discuss stability, defined in the sense of Lyapunov, and then present a method to test whether a system is stable (the "indirect method").

7.3.1 Stability in the Sense of Lyapunov

We consider a generic control system, with a state vector $x \in \mathbb{R}^n$, inputs $u \in \mathbb{R}^m$, with a control law $u = k(x)$, so that we have the closed-loop dynamics

$$\dot{x} = f(x, u) = f(x, k(x)) =: \bar{f}(x) \tag{7.30}$$

and our goal is to discuss *stability* of the system (7.30).

In simple English, the system is stable near some value if small perturbations only have small consequences; a system is not stable if small perturbations can have large consequences. We can make this intuitive definition concrete using the language of Lyapunov; and for this we require the concept of equilibria from (7.2). We will consider only closed-loop systems, with equilibria x^* so that $\bar{f}(x^*) = 0$, as defined in (7.5).

Now that we've defined systems and their equilibria, we can make precise what we mean by stability.

Definition 7.1 Consider the system $\dot{x} = \bar{f}(x)$, and let x^* be a specific equilibrium of the system.

- The equilibrium x^* is a *stable equilibrium* if, for any $\epsilon > 0$, there exists a $\delta > 0$ so that, if $\|x(0) - x^*\| < \delta$, then $\|x(t) - x^*\| < \epsilon$ for all times $t \geq 0$.
- The equilibrium x^* is an *unstable equilibrium* if it is not stable.

The definition of stability, at first glance, seems complex, but makes sense after unpacking it. First some high-level points: A very important note is that we're not defining whether the *system* is stable, rather whether an *equilibrium* is stable. A single system may have multiple equilibria, some of which are stable, and some of which are unstable. The second point is that an equilibrium can be stable, or unstable, but nothing else: specifically, there is no concept of "borderline" stable, or "marginally" stable in Lyapunov.

Let's unpack the definition:

- The statement $\|x(0) - x^*\| < \delta$ means that our initial condition $(x(0))$ is no farther than δ away from the equilibrium.
- The statement $\|x(t) - x^*\| < \epsilon$ means that our state at time t $(x(t))$ remains no farther than ϵ away from the equilibrium.

We can imagine both of these as spheres (in n-dimensional state space) centered at the equilibrium x^*.

Thus, for an equilibrium to be stable: if we want the system to remain 'close' to the equilibrium, and specifically never move as far as ϵ away from the equilibrium; we have to be able to specify a ball of radius δ, so that if we start within the δ ball, we never leave the ϵ ball. Note that we cannot specify an empty ball for the initial condition, because δ must be strictly positive.

One can think of this as defining a 'danger radius' ϵ: we want the state to remain within the danger radius for all future times. For this danger radius, can we define a 'safe starting radius' δ, so that any starting condition in the safe starting radius never exceeds the danger radius? If we can find such a δ for any ϵ, then the equilibrium is stable.

We notice that, for an unstable equilibrium, there must exist some value $\tilde{x}$ which is arbitrarily close to x^*, so that if we start the system at $x(0) = \tilde{x}$ the system moves farther away from the equilibrium. An important note here is that not all initial conditions have to have this property—an equilibrium is unstable even if there are some initial conditions which remain close to the equilibrium, as long as there are also some which would push us further away.

Thus, again in English—a stable equilibrium is a system state so that, if the system starts sufficiently close to that equilibrium ($\|x(0) - x^*\| < \delta$) then it remains close to that equilibrium for all time ($\|x(t) - x^*\| < \epsilon$). An unstable equilibrium is one for which it is possible to start close the equilibrium, but have the system move far away nonetheless. The simplest example might be a rigid pendulum—this has two equilibria, one with the pendulum hanging vertically down, and another with the pendulum standing vertically upright. The upright equilibrium is unstable: if we start ever so slightly off of the vertical, the pendulum will inevitably fall away. However, the downward equilibrium is stable—if we start near the equilibrium, the pendulum may oscillate, but will remain near the equilibrium forever.

Note that a stable equilibrium need not be *attractive*, i.e. we don't require that $x(t) \to x^*$ as $t \to \infty$. Indeed, for a hypothetically frictionless pendulum, it would continue oscillating about the stable downward equilibrium for all time, without ever coming to rest there.

Now that we have a clean definition of what we mean by *stable* we turn our attention to how to test whether something is, in fact, stable. We note that the definition is not particularly useful in this regard; the definition only tells us which properties an equilibrium must have. Consider a first-order system, $\dot{x} = -\frac{1}{\tau}x$ for $\tau > 0$ a time constant. Of course, we have already argued that for $\tau > 0$ this is well-behaved system, our objective is to show that this is also *stable in the sense of Lyapunov*.

We know the solution to this system immediately:

$$x(t) = e^{-\frac{t}{\tau}}x(0) \tag{7.31}$$

Taking the magnitude of the above, and noting that $e^{-\frac{t}{\tau}} > 0$, yields

$$\|x(t)\| = \left\| e^{-\frac{t}{\tau}}x(0) \right\| = e^{-\frac{t}{\tau}} \|x(0)\| \tag{7.32}$$

Thus, for any value $\epsilon > 0$, we can choose (say) $\delta = \frac{1}{2}\epsilon$, so that for any $x(0)$ with $\|x(0)\| \leq \delta$

$$\|x(t)\| = e^{-\frac{t}{\tau}} \|x(0)\| \tag{7.33a}$$

$$\leq e^{-\frac{t}{\tau}}\delta \tag{7.33b}$$

$$\leq \delta = \tfrac{1}{2}\epsilon < \epsilon \tag{7.33c}$$

because $e^{-\frac{t}{\tau}} \leq 1$ for any $t \geq 0$. Thus we have proven that for every $\epsilon > 0$ there exists a $\delta > 0$ (etc. etc.), so that by Definition 7.1 we can conclude this equilibrium is stable in the sense of Lyapunov.

7.3.2 Testing for Stability: Lyapunov's Indirect Method

From the preceding, we have a clear definition of *stability*, and we have seen an example of how we can test for stability if we can solve the differential equation of the system. However, most systems do not readily admit a solution to their ODEs, or such solutions are too complex to directly inspect. We thus require a more easily applied method for testing for stability.

Lyapunov provides two methods for establishing that an equilibrium is stable: the "indirect" method (sometimes called the 'second' method) and the "direct" method (sometimes called the 'first' method). We will only consider the indirect method here—this is easier to use; however the direct method may provide richer information.

We consider the general closed-loop system $\dot{x} = \bar{f}(x)$, where we are considering equilibrium x^* where we assume that $x^* = 0$ for simplicity (as in Sect. 7.1.1, we can always redefine the states so that this is possible). Consider the first order approximation of this system, where $z \approx x$ and

$$\dot{z} = Az, \qquad \text{where } A := \left.\frac{\partial \bar{f}}{\partial x}\right|_{x=x^*} \tag{7.34}$$

By this choice, we note that (by a Taylor series expansion) we have that the difference between the general system x and the first order approximation grows as

$$\dot{x} - \dot{z} = \bar{f}(x) - Az = A(x - z) + \text{HOT}(x) \tag{7.35}$$

wherein $\text{HOT}(x)$ are higher-order terms in x (that is, things that go to zero faster than x goes zero). In other words, when $x(0)$ is close to zero, and $z(0) = x(0)$, then $z(t)$ will remain close to $x(t)$ as long as they remain near zero. Now we state the indirect method.

Theorem 7.1 *Lyapunov's indirect method: Consider the system $\dot{x} = \bar{f}(x)$, with an equilibrium at $x^* = 0$, and let $\dot{z} = Az$ be a first-order approximation thereof, with $A = \frac{\partial \bar{f}}{\partial x}$. Then:*

- *If $\dot{z} = Az$ is exponentially stable (i.e. if all eigenvalues of A have negative real parts), then $\dot{x} = \bar{f}(x)$ is stable in the sense of Lyapunov at $x^* = 0$.*
- *If A has at least one exponentially unstable eigenvalue (i.e. one eigenvalue of A has a strictly positive real part), then $\dot{x} = \bar{f}(x)$ is unstable in the sense of Lyapunov at $x^* = 0$.*
- *If A has no eigenvalues with positive real parts, but at least one eigenvalue which is purely imaginary (i.e. has zero real part), then $\dot{x} = \bar{f}(x)$ may be either stable or unstable.*

Thus, we can test for the stability of $\dot{x} = \bar{f}(x)$ indirectly, by computing the eigenvalues of the corresponding linear system's dynamics matrix. We also note that the indirect method can sometimes fail to reach a conclusion (if there is an eigenvalue with zero real part).

Let us first understand why analyzing the linearized system allows for conclusions on the nonlinear system. We note that, if we start sufficiently close to the equilibrium (i.e. $\|x(0)\| \ll 1$), then (at least initially) $\dot{x} \approx \dot{z}$. If we can now show that $z \to 0$, then (because x remains close to z as long as they're small), then $x \to 0$ because the linearization remains accurate, and we can conclude the equilibrium $x^* = 0$ is stable. If, however, we can show that z will "blow up", then x will (at least initially) grow with z, and the equilibrium is unstable in the sense of Lyapunov. Of course, as x grows, the linearization may become a poor approximation, and x may eventually evolve very differently from z.

Next up, we try to understand the bizarre fact that properties from linear algebra (eigenvalues) allow us to determine the properties of a dynamic system. To see why this is true, we look at some examples—consider the case where A is diagonal, so that

$$\dot{z} = Az, \qquad A = \begin{bmatrix} \lambda_1 & 0 & \dots & 0 \\ 0 & \lambda_2 & \dots & 0 \\ \vdots & \vdots & \ddots & \vdots \\ 0 & 0 & \dots & \lambda_n \end{bmatrix} = \operatorname{diag}(\lambda_1, \lambda_2, \dots, \lambda_n) \tag{7.36}$$

Clearly, the eigenvalues of A are simply λ_i, and we can look at the evolution of each component of z independently: $\dot{z}_i = \lambda_i z_i$. We can thus solve exactly the state z at any time in the future: $z_i(t) = e^{\lambda_i t} z_i(0)$. Thus, we can immediately say that $z \to 0$ for any $z(0)$ if, and only if, all λ_i are negative. If, on the other hand, any one of λ_i is positive, then at least one z_i will grow without bound, and $\|z(t)\| \to \infty$ as $t \to \infty$.

Of course, most interesting systems do not neatly decouple as above, and generally the A matrix is dense and has complex (rather than purely real) eigenvalues. Let us make the assumption that the matrix can be made diagonal, so that there exists some matrix $T \in \mathbb{C}^{n \times n}$ so that:

$$TAT^{-1} = \operatorname{diag}(\lambda_1, \lambda_2, \dots, \lambda_n) \tag{7.37}$$

In this case, we can look at the dynamic system with state $y := Tz$ so that $z = T^{-1}y$. For this system, we have:

$$\dot{y} = T\dot{z} = TAz = TAT^{-1}y = \operatorname{diag}(\lambda_1, \lambda_2, \dots, \lambda_n)\, y \tag{7.38}$$

and we have that $\dot{y}_i = \lambda_i y_i$, and $y_i(t) = e^{\lambda_i t} y_i(0)$. Again, if $\operatorname{Re}(\lambda_i) < 0$ we have that $y_i \to 0$ as $t \to 0$. Noting that z goes to zero if and only if y goes to zero; we can conclude that if y goes to zero (or diverges) then z will go to zero (or diverge), and by our linearization argument therefore x will go to zero (or, at least initially, diverge).

Not all matrices, unfortunately, can be diagonalized in this way. The most general case requires the "Jordan decomposition", a proper treatment of which is beyond the scope of this book.[5] For such cases, though, it can be shown without too much effort that z will converge to zero only if all eigenvalues have negative real parts, and will diverge if there is at least one positive real eigenvalue.

7.4 Other General Control Approaches

Now that we have an idea of what control may be, and how to reason concretely about conditions such as stability, it may be interesting to consider alternative control approaches (that are more sophisticated than that of Sect. 7.2). Such a discussion would be well beyond the scope of this book, but below are some popular techniques that may be of interest. A common thread is that they are all heavily model-driven—in aerospace applications especially, the system model tends to be well understood; this in turn allows for sophisticated and powerful control techniques.

LQR (Linear Quadratic Regulator): An approach for synthesizing a linear feedback controller. The control design problem is framed as choosing the weights for a quadratic cost function (weighing errors in state and input). The result is an *optimal controller*, although the optimality is usually not central (rather, it is easy to understand, design for, and deploy). LQR controllers are very widely used, and have excellent robustness properties. Good references are Anderson and Moore (2007) and Bertsekas (2005).

MPC (Model Predictive Control): An approach which solves an optimal control problem in real time (i.e., at every time step where a control input must be computed, a new optimization problem is solved). This allows to explicitly consider the future desired trajectory, for example. This is typically much more computationally expensive than LQR (which only requires a matrix multiplication in real time), but can be more flexible. Most commonly, an MPC controller is applied to a linearized model, with a quadratic cost function (similar to the cost function in LQR), but with the additional ability to explicitly encode constraints (such as actuator limits, or safety constraints on the states). A good reference is Borrelli et al. (2017).

Robust control: Here one typically tries to find the feedback control which minimizes the *worst-case* performance, such as $\mathcal{H}_\infty$ control. Outside of the worst-case, the performance tends to be worse than something like LQR though. See, for example, Dullerud and Paganini (2013).

Adaptive, data driven, or learning control: This is a very broad category, that includes many things. There are many adaptive control approaches, where the controller is meant to adapt to unknown parameters (for example, the unknown total mass of the aircraft). This includes techniques like the 'MIT rule' (Åström

[5] For exhaustive info, see the book by Bernstein (2009), or a brief overview in the Wikipedia entry https://en.wikipedia.org/wiki/Jordan_normal_form. As an aside, the Bernstein book is great to leave on your desk to intimidate others.

and Wittenmark 1995), or more modern ideas like $\mathcal{L}_1$ control (Cao and Hovakimyan 2008). More recently, techniques using machine learning have shown great promise, especially for closing the loop around high-dimensional sensor data (such as vision systems)—one example is Kaufmann et al. (2023).

Geometric control: With geometric control, the controller is designed to exploit the mathematical ('geometric') structure of the system dynamics, to create a closed-loop system with better performance for states far from the nominal. An example would be an attitude control system, wherein the control design relies on properties of orientations. A good overview of geometric mechanics is Lee et al. (2017), with some particular attitude control examples given in Mueller (2018).

7.5 A Quadcopter Control Architecture

Next, we apply the above control theory to our guiding example system—and we develop a cascaded control strategy for a quadcopter. The idea is to break the quadcopter up into a collection of simple systems, apply simple control strategies to these systems, so that the quadcopter achieves the desired behavior.

We will construct the cascaded controller as follows: the lowest level control will control the vehicle's angular velocity, p, q, and r, using the torques n_1, n_2, and n_3 as inputs. The next level up will control the quadcopter's attitude, here expressed as the Euler angles ψ, θ, and ϕ, using the angular velocities as inputs. The final layer will control the vehicle's position/velocity, using the total thrust c_Σ and attitude as inputs.

We will assume that we have perfect knowledge of the system's state. In practice, we never have such perfect knowledge, and instead rely on state estimators as designed in Chap. 5.

The scheme is illustrated in Fig. 7.5.

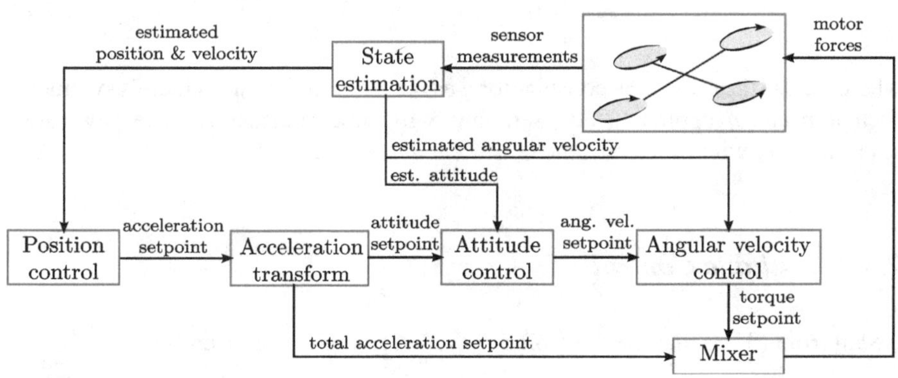

Fig. 7.5 A typical cascaded controller structure for a quadcopter

7.5.1 Angular Velocity Control

Recall the dynamics of the vehicle's angular velocity:

$$\begin{bmatrix} J_{xx}\dot{p} \\ J_{xx}\dot{q} \\ J_{zz}\dot{r} \end{bmatrix} = \begin{bmatrix} n_1 \\ n_2 \\ n_3 \end{bmatrix} - \begin{bmatrix} (J_{xx} - J_{zz})\, qr \\ (J_{zz} - J_{xx})\, pr \\ 0 \end{bmatrix} \tag{7.39}$$

We will do the analysis for the roll axis, with p as state, and then generalize to the other states. Define u_p as

$$u_p := \frac{1}{J_{xx}} (n_1 - (J_{xx} - J_{zz})\, qr) \tag{7.40}$$

so that we get the simple $\dot{p} = u_p$.

Let p_s be the roll rate setpoint. Selecting $u_p = \frac{1}{\tau_p}(p_s - p)$ will cause the vehicle roll-rate to converge to p_s with time constant τ_p. This is achieved for

$$n_1 = J_{xx} u_p + (J_{xx} - J_{zz})\, qr \tag{7.41a}$$

$$= \frac{J_{xx}}{\tau_p}(p_s - p) + (J_{xx} - J_{zz})\, qr \tag{7.41b}$$

Note that this is a form of *feedback linearization*—by suitably redefining an input, we have managed to hide the system's inherent non-linearity, without resorting to approximations/linearization.

Similar arguments yield

$$n_2 = \frac{J_{xx}}{\tau_p}(q_s - q) + (J_{zz} - J_{xx})\, pr \tag{7.42}$$

$$n_3 = \frac{J_{zz}}{\tau_r}(r_s - r) \tag{7.43}$$

where we use the same time constant for q as for p because of the vehicle's symmetry, so that p and q approach their setpoints with time constant τ_p. The yaw rate r approaches r_s with time constant τ_r.

7.5.2 Attitude Control

Recall from (3.23) that the dynamics of the Euler angles are given by

$$\frac{\mathrm{d}}{\mathrm{d}t}\begin{bmatrix}\phi\\ \theta\\ \psi\end{bmatrix} = \begin{bmatrix}1 & \sin\phi\tan\theta & \cos\phi\tan\theta\\ 0 & \cos\phi & -\sin\phi\\ 0 & \sin\phi/\cos\theta & \cos\phi/\cos\theta\end{bmatrix}\begin{bmatrix}p\\ q\\ r\end{bmatrix} \tag{7.44}$$

or, for small angles,

$$\dot{\phi} \approx p \tag{7.45}$$

$$\dot{\theta} \approx q \tag{7.46}$$

$$\dot{\psi} \approx r \tag{7.47}$$

We use p_s as the desired roll rate (an input here), and let ϕ_s be the roll angle setpoint. Then, setting $u_\phi = \frac{1}{\tau_\phi}(\phi_s - \phi)$ will cause the roll angle to approach the setpoint with time constant τ_ϕ.

We can use a similar approach for θ and ϕ, where again by symmetry we use the same time constant for ϕ as for θ, and use the time constant τ_ψ for the yaw angle ψ.

7.5.3 Position Control

We will consider one direction in inertial space, with position p and velocity v (with $\dot{p} = v$), and we will treat the acceleration as input, so that $\dot{v} = u$. Following the argument of Sect. 7.2.2, we choose a natural frequency ω_n and a damping ratio ζ, and choose the input $u = -2\zeta\omega_n v - \omega_n^2(p_s - p)$, where p_s is the position setpoint.

7.5.4 Acceleration Transform

The acceleration transform takes the acceleration setpoint, and outputs an attitude setpoint and a total acceleration setpoint. This is a transformation, not a controller, since there is no concept of feed-back. Given an acceleration setpoint, in an earth-fixed coordinate system, $\left[\underline{a_{des}}\right]^E$, we transform this to a total acceleration setpoint, and attitude setpoint, as follows. Recall that the acceleration (a_1, a_2, a_3) of a quadcopter is related to its attitude (as transformation matrix $[T]^{EB}$ and the total thrust c_Σ as

$$m^B\begin{bmatrix}a_1\\ a_2\\ a_3\end{bmatrix} = c_\Sigma\,[T]^{EB}\begin{bmatrix}0\\ 0\\ 1\end{bmatrix} + m^B\begin{bmatrix}0\\ 0\\ -\left\|\underline{g}\right\|\end{bmatrix} \tag{7.48}$$

We note that we can immediately solve for c_Σ, since the transformation preserves the norm of a vector:

$$c_\Sigma = m^B \left\| \begin{bmatrix} a_1 \\ a_2 \\ a_3 + \left\| \underline{g} \right\| \end{bmatrix} \right\| = m^B \sqrt{a_1^2 + a_2^2 + \left(a_3 + \left\| \underline{g} \right\|\right)^2} \tag{7.49}$$

We can now solve for the Euler angles that describe the transformation matrix,

$$[T]^{EB} = \begin{bmatrix} \mathrm{c}\psi\mathrm{c}\theta & \mathrm{c}\psi\mathrm{s}\theta\mathrm{s}\phi - \mathrm{s}\psi\mathrm{c}\phi & \mathrm{c}\psi\mathrm{s}\theta\mathrm{c}\phi - \mathrm{s}\psi\mathrm{s}\phi \\ \mathrm{s}\psi\mathrm{c}\theta & \mathrm{s}\psi\mathrm{s}\theta\mathrm{s}\phi + \mathrm{c}\psi\mathrm{c}\phi & \mathrm{s}\psi\mathrm{s}\theta\mathrm{c}\phi - \mathrm{c}\psi\mathrm{s}\phi \\ -\mathrm{s}\theta & \mathrm{c}\theta\mathrm{s}\phi & \mathrm{c}\theta\mathrm{c}\phi \end{bmatrix} \tag{7.50}$$

or, for ϕ, θ, and ψ small

$$m^B \begin{bmatrix} a_1 \\ a_2 \\ a_3 \end{bmatrix} \approx c_\Sigma \begin{bmatrix} \theta \\ -\phi \\ 1 \end{bmatrix} + m^B \begin{bmatrix} 0 \\ 0 \\ -\left\| \underline{g} \right\| \end{bmatrix} \tag{7.51}$$

so that we can solve the angle setpoints from the acceleration values as

$$\theta_s := \frac{a_1}{\sqrt{a_1^2 + a_2^2 + \left(a_3 + \left\| \underline{g} \right\|\right)^2}} \tag{7.52a}$$

$$\phi_s := -\frac{a_2}{\sqrt{a_1^2 + a_2^2 + \left(a_3 + \left\| \underline{g} \right\|\right)^2}} \tag{7.52b}$$

Thus we have a workable control strategy for a quadcopter. Of course, for demanding applications it may be worth spending the effort to design a better, more integrated controller. One may also prefer to not use Euler angles at all in control (you really shouldn't use them in control if there is *any* chance that the angles deviate from being 'small').

7.6 Trajectory Generation and Tracking

We briefly discussed feedforward signals in Sect. 7.2.3, we now revisit this in slightly more detail and consider the problem of *trajectory generation*, i.e., how do we get our system from one state to another state. For example, consider the problem of a quadcopter initially at point S, which we want to move to a final position F—what are the inputs we need to generate so that the system arrives at that point?

More generally, consider a system with state vector x, input u, which satisfies the differential equation $\dot{x} = f(x, u)$. For our problem, we assume that we have a given initial state, x_S, so that at the initial time t_s we have $x(t_s) = x_s$; and we are given a target final state x_F, which we want the system to reach at time t_F, so that

$x(t_F) = x_F$. The trajectory generation problem is then to find a function $u_{\text{ref}}(t)$ (i.e., a sequence of inputs over time) so that the following differential equation is satisfied:

$$\dot{x}_{\text{ref}}(t) = f(x_{\text{ref}}(t), u_{\text{ref}}(t)) \qquad \text{for } t \in [t_S, t_F] \tag{7.53}$$

$$x_{\text{ref}}(t_S) = x_S \tag{7.54}$$

$$x_{\text{ref}}(t_F) = x_F \tag{7.55}$$

This is typically an underdetermined problem (there are many different routes that I can take from my house to the grocery store, each of which satisfies the above expression). Thus, it is common to pose it as an *optimal control problem*, where we seek to not only solve the above differential equation, but also minimize some specified cost (such as fuel use along the trajectory). Typically, specifying a requirement of optimality is enough to make the solution unique.[6]

This pair of state and input functions (x_{ref} and u_{ref}) is then a *reference trajectory* that we can follow, with the idea that if the true system does start at $x(t_0) = x_S$, and we simply apply $u_{\text{ref}}(t)$ over $t \in [t_S, t_F]$, then at time t_F the system arrives at the final goal $x(t_f) = x_F$. Of course, in practice, we never start exactly at x_S, nor is our dynamics model (7.53) exact (for example, there will be disturbances acting on our system).

We can thus investigate the deviation of the system from the reference, and define two deviation signals (for the state x_Δ and the input u_Δ), so that

$$x(t) = x_{\text{ref}}(t) + x_\Delta(t) \tag{7.56}$$

$$u(t) = u_{\text{ref}}(t) + u_\Delta(t) \tag{7.57}$$

Rearranging, then taking the derivative of (7.56); and substituting the system dynamics gives

$$\dot{x}_\Delta = \dot{x} - \dot{x}_{\text{ref}} \tag{7.58a}$$

$$= f(x_{\text{ref}} + x_\Delta, u_{\text{ref}} + u_\Delta) - f(x_{\text{ref}}, u_{\text{ref}}) \tag{7.58b}$$

$$\approx f(x_{\text{ref}}, u_{\text{ref}}) + \frac{\partial f}{\partial x} x_\Delta + \frac{\partial f}{\partial u} u_\Delta - f(x_{\text{ref}}, u_{\text{ref}}) \tag{7.58c}$$

$$= A(t)x_\Delta + B(t)u_\Delta \tag{7.58d}$$

where we did a first-order expansion of the dynamics, and introduced the matrices:

$$A(t) := \left.\frac{\partial f}{\partial x}\right|_{x_{\text{ref}}(t), u_{\text{ref}}(t)} \tag{7.59}$$

$$B(t) := \left.\frac{\partial f}{\partial u}\right|_{x_{\text{ref}}(t), u_{\text{ref}}(t)} \tag{7.60}$$

[6] Typically at least locally, though not always, for more infomation see, e.g., the Pontryagin Minimum Principle in Liberzon (2011).

We thus have that the system tracking error x_Δ evolves (approximately) as a linear, first-order dynamic system. We can design the deviation input u_Δ through some feedback policy, perhaps a linear policy so that $u_\Delta(t) = -K(t)x_\Delta(t)$ so that $x_\Delta \to 0$, and thus $x(t) \to x_{\text{ref}}(t)$. We then apply the control input as $u(t) = u_{\text{ref}}(t) + u_\Delta(t)$, thus consisting of a feedforward and a feedback term. Because this is a time-varying system, designing such a controller may be more challenging than what we've discussed in this book. The most natural tool to use may be an LQR controller, see e.g. Anderson and Moore (2007).

Notice that this is a powerful idea, and we can push it farther. For example, solving a reference input for the exact dynamics of (7.53) may be prohibitively difficult. We may thus settle for an approximate solution, so that

$$\dot{x}_{\text{ref}}(t) \approx f(x_{\text{ref}}(t), u_{\text{ref}}(t)) \tag{7.61}$$

and we treat the approximation error simply as a disturbance that our feedback controller should reject.

This may take a few different forms. For the sake of a concrete example, consider trying to fly a quadcopter from one hover condition to another, with a horizontal displacement between them. Solving the differential equation exactly would require solving the four motor forces over the time of the trajectory, and satisfying the challenging, nonlinear dynamics of the vehicle. We imagine a series of possible approximations, starting with an extremely crude approach: Imagine a reference trajectory which is much simpler, split into two distinct phases: for $t < t_S$, we simply apply the hover controller at the initial position at point S. When $t \geq t_S$ we apply the hover controller at point F. This is clearly very crude approach, and may work acceptably when the distance between S and F are sufficiently close. However, when the points are far away, this strategy will not work well.

One step of sophistication higher, we may consider splitting the flight into three segments. We still use the hover controller: for $t < t_S$ we command the system to hover at point S, and for $t > t_F$ we hover at point F. However, for the intermediate time, we simply command the vehicle to fly at constant velocity between the two points. Clearly, this reference is not *feasible*, in the sense that it does not satisfy the differential equation (7.53), as a system cannot change its velocity instantaneously. However, this discontinuity is less dramatic than if we simply do two hover controllers, and we expect the system to be able to track it reasonably well.

If we continue, we may design a trajectory that has perhaps five phases, starting at rest, then accelerating at constant acceleration until a fixed speed, maintaining that speed, then decelerating at constant acceleration to end at rest at the target location. This reduces the discontinuity, but there is still a discontinuity—for a quadcopter to change its acceleration requires a change of orientation, which can of course not happen instantaneously.

Systematic approaches to this problem include the Minimum Principle (see e.g. Liberzon 2011), which turns an optimal trajectory generation problem into a partial differential equation (but leaves you with the challenging task of solving a partial differential equation!). Another approach immediately discretizes time, so that the problem is not to solve a function $u_{\text{ref}}(t)$, but instead a finite number of discrete-time inputs $u_{\text{ref}}(k\Delta t)$ at the integer time steps k—this is typically easier as we have very powerful techniques for solving optimization problems over a discrete set of variables (as opposed to an infinite-dimensional function). Under certain conditions, the optimization problem can be made *convex*, in which case extremely powerful numerical techniques exist to solve the optimization problems (which is often the foundation for Model Predictive Control, see e.g. Borrelli et al. 2017).

Problems for Chapter 7

Problem 7.1 (*5 points*) Consider a linear system with state $x \in \mathbb{R}^n$, and input $u \in \mathbb{R}^m$, evolving according to the dynamics $\dot{x} = Ax + Bu$ with A and B constant matrices of appropriate dimension. Show that this system always has at least one equilibrium, and either has only one, or an infinite number of equilibria.

Problem 7.2 (*6 points*) Consider the first-order system $\dot{x} = u$, with initial condition $x(0) = 1$. Let the reference be a simple sinusoid, $x_s = \sin(t)$; and implement a proportional controller with time constant be $\tau = 1$. Implement a naive controller without feedforward, and a controller with feedforward, and simulate the two responses. Make a single graph, with three lines: the reference state x_s, and the system response x with and without the feedforward term. How do they compare? What happens if you include the feedforward, but not the feedback?

Problem 7.3 (*6 points*) Consider a double integrator, with a reference trajectory p_s which is not constant. How do we need to modify our controller of Sect. 7.2.2 to track this trajectory well? Again, simulate such a system tracking the desired position trajectory $p_s = \sin(t)$ with a controller with natural frequency $\omega_n = 1\,\text{rad}\,\text{s}^{-1}$ and damping ratio $\zeta = 0.7$.

Problem 7.4 (*4 points*) Show that the cascaded approach to control a second-order system (Sect. 7.2.4) results in an equivalent natural frequency $\zeta = \frac{1}{2}\sqrt{\frac{\tau_p}{\tau_v}}$, and natural frequency $\omega_n = \frac{1}{\sqrt{\tau_p \tau_v}}$.

Problem 7.5 (*6 points*) Implement a triple integrator in simulation, and apply the cascaded control of Sect. 7.2.4. Show, using simulations, that the system behaves well if $\tau_1 \gg \tau_2 \gg \tau_3$, i.e. if cascade of systems is much faster the deeper into the cascade we are. Show, using simulations, that the closed-loop system remains stable for any choices of time constants as long as $\tau_3 < \tau_1$. Show, using simulations, if $\tau_3 > \tau_1$, the system becomes unstable, and the error eventually will grow arbitrarily large.

Problem 7.6 (*8 points*) Use the indirect method to predict stability of a cascaded controller for a triple integrator. Fix $\tau_1 = 4$, $\tau_2 = 2$, and vary τ_3. For which values of τ_3 do you predict the system to transition from stable to unstable? Validate this using your simulation from Problem 7.5.

Problem 7.7 (*6 points*) Consider a dynamic system described by $\dot{x} = f(x, u)$ where $x = (x_1, x_2) \in \mathbb{R}^2$ is the system state, and $u \in \mathbb{R}$ is the input. Given is

$$f(x, u) = \begin{bmatrix} x_2 + u \\ x_1 \end{bmatrix} \tag{7.62}$$

Find all equilibria for this system.

Problem 7.8 (*8 points*) Consider a dynamic system described by $\dot{x} = f(x, u)$ where $x = (x_1, x_2) \in \mathbb{R}^2$ is the system state, and $u \in \mathbb{R}$ is the input. Given is

$$f(x, u) = \begin{bmatrix} x_2 \left(x_1^2 + 1\right) \\ -x_1 + u \end{bmatrix} \tag{7.63}$$

Find all equilibria for this system.

Problem 7.9 (*6 points*) Consider the system $\dot{x} = f(x)$, with $x = (x_1, x_2)$, and

$$f(x) = \begin{bmatrix} (1 + x_1) \cos(x_2) \\ (1 - x_1) x_2 \end{bmatrix}$$

Compute all equilibria for this system.

Problem 7.10 (*4 points*) Consider a system with state $x \in \mathbb{R}$, which evolves according to the differential equation $\dot{x} = f(x) = x^2 + 1$. Compute any/all equilibria of this system.

Problem 7.11 (*8 points*) Consider a dynamic system described by $\dot{x} = f(x, u)$ where $x \in \mathbb{R}$ is the system state, and $u \in \mathbb{R}$ is the input. Given is

$$f(x, u) = ux^3 + 2x + u \tag{7.64}$$

Note that $x^* = 0$, $u^* = 0$ is an equilibrium of this system

The system's control input is given by the control law $u = -kx$. For which values of k will the closed-loop system be stable (in the sense of Lyapunov) at the equilibrium?

Problem 7.12 (*8 points*) Consider a system with state $x \in \mathbb{R}$, with dynamics

$$\frac{\mathrm{d}}{\mathrm{d}t} x = 5x^2 + x$$

Use Lyapunov's indirect method to characterize the stability all equilibria of the system.

Problem 7.13 (*10 points*) Consider the design of an attitude controller for the quadcopter's pitch angle, characterized by two time constants: τ_q as the time constant for the angular velocity, and τ_θ as the time constant for the vehicle angle. These are used as follows:

$$q_{\text{cmd}} := -\frac{1}{\tau_\theta}\left(\hat{\theta} - \theta_{\text{cmd}}\right); \qquad \dot{q}_{\text{cmd}} := -\frac{1}{\tau_q}\left(\hat{q} - q_{\text{cmd}}\right); \qquad n_2 = J_{xx}\,\dot{q}_{\text{cmd}} \tag{7.65}$$

For the remainder of this problem, set $\theta_{\text{cmd}} = 0$, and assume that we have a perfect state estimator, i.e. $\hat{\theta} = \theta$ and $\hat{q} = q$. Use $J_{xx} = 16 \times 10^{-6}\text{kgm}^2$.

You want to choose the values τ_i to be fast, but not lead to actuator saturation (due to input values that are too large). From extensive modelling you know that, in normal operation and due to "normal disturbances", the true angle θ should never exceed $\bar{\theta}$ in magnitude, and the true angular velocity should never exceed $\bar{q}$ in magnitude (note that these limits may be achieved simultaneously).

Using the relationship $\tau_\theta = 2\tau_q$, choose the time constant τ_θ so that the response is as fast as possible, while guaranteeing that if

$$|\theta| \le \bar{\theta} = \frac{\pi}{6}\text{rad and } |q| \le \bar{q} = 20 \text{ rad s}^{-1} \text{ then} \tag{7.66}$$

$$|n_2| \le \bar{n}_2 = 3.9 \times 10^{-3}\text{N m} \tag{7.67}$$

Problem 7.14 (*6 points*) In this problem we will analyze nonlinear systems and their first-order approximations. For each system i, we consider the equilibrium $x_i^* = 0$. Consider three systems x_i and their corresponding first-order approximations z_i:

$$\dot{x}_1 = x_1 + x_1^2 \qquad \dot{z}_1 = z_1 \tag{7.68}$$

$$\dot{x}_2 = -\tan(x_2) \qquad \dot{z}_2 = -z_2 \tag{7.69}$$

$$\dot{x}_3 = 1 - \cos(x_3) \qquad \dot{z}_3 = 0 \tag{7.70}$$

For each of these systems, compute the range for $|x_i|$ so that, if $z_i = x_i$ then $\frac{\dot{z}_i}{\dot{x}_i} \in [0.8, 1.2]$. *Hint: you may solve this numerically, you do not need to provide a closed-form solution.*

How does this relate to your understanding of testing for stability using the indirect method?

Problem 7.15 (*8 points*) Consider a quadruple integrator, i.e. a system where:

$$\frac{d}{dt}\begin{bmatrix} x_1 \\ x_2 \\ x_3 \\ x_4 \end{bmatrix} = \begin{bmatrix} x_2 \\ x_3 \\ x_4 \\ u \end{bmatrix} \tag{7.71}$$

with x_i the states, and u the scalar input.

We design a nested controller for the system, with the following control structure:

$$x_{2,\mathrm{cmd}} = -\frac{1}{\tau_1}\left(x_1 - x_{1,\mathrm{cmd}}\right) \tag{7.72}$$

$$x_{3,\mathrm{cmd}} = -\frac{1}{\tau_2}\left(x_2 - x_{2,\mathrm{cmd}}\right) \tag{7.73}$$

$$x_{4,\mathrm{cmd}} = -\frac{1}{\tau_3}\left(x_3 - x_{3,\mathrm{cmd}}\right) \tag{7.74}$$

$$u = -\frac{1}{\tau_4}\left(x_4 - x_{4,\mathrm{cmd}}\right) \tag{7.75}$$

Compute the possible values for ρ, so that the closed-loop system is stable in the sense of Lyapunov, where

$$\tau_1 = 1\mathrm{s} \qquad \text{and} \qquad \tau_{i+1} = \frac{\tau_i}{\rho}. \tag{7.76}$$

Hint—a numerical solution will suffice, you do not need to compute an analytical solution.

Problem 7.16 (*12 points*) Consider closed-loop systems of the form $\dot{x} = f(x)$, with $x = (x_1, x_2) \in \mathbb{R}^2$ (where the control input has been replaced by an appropriate control law) so that each system has an equilibrium at $x = 0$ (the state of interest).

For each of the following systems, use Lyapunov's indirect method to comment on stability in the sense of Lyapunov. Clearly motivate your answers.

(a) System: $f(x) = \begin{bmatrix} -x_1 + x_1^3 + x_2 \\ -x_2 \end{bmatrix}$

(b) System: $f(x) = \begin{cases} \begin{bmatrix} -x_1 \\ -2x_2 \end{bmatrix} & \text{if } |x_1| > 10^{-3} \\[2ex] \begin{bmatrix} x_1 \\ -2x_2 \end{bmatrix} & \text{else.} \end{cases}$

Problem 7.17 (*10 points*) We have learnt the *indirect* method of Lyapunov. In this problem we will build some intuition towards the *direct* method. The direct method is usually much harder to apply, and has a variety of technical details associated which we won't go into in this book. However, the advantage is that, unlike the indirect method, it is possible to prove that a system will stay near the equilibrium *even if the initial condition is not near the equilibrium*. Recall that the indirect method only allowed us to say that the system will remain near the equilibrium as long as we start sufficiently close to the equilibrium, but it didn't help us answer the question "how close to the equilibrium?", or more pointedly, "how near is near enough?".

Consider a spring-mass-damper system, where a cart of mass m moves without friction along a horizontal surface. The cart is at a position p with speed v. The cart is attached to a wall with a spring (with the spring force $f_s = -kp$, where k is the spring constant) and a damper (with damping force $f_d = -cv$, where c is the damping constant). Given is that $k > 0$ and $c > 0$. We now have the system dynamics:

$$\ddot{p} = \dot{v} = \frac{1}{m}(-kp - cv) \tag{7.77}$$

We already know intuitively that the cart will return to $p = v = 0$ no matter where it starts; we now prove this using an argument similar to what is used for the *direct method*.

The mechanical energy of the cart (consisting of the kinetic energy of its motion, and the potential energy in the spring) is given by $E_m = \frac{1}{2}mv^2 + \frac{1}{2}kp^2$.

(a) Show that the system only has one equilibrium at $p^* = v^* = 0$.
(b) Using the *indirect method*, show that the equilibrium is stable.
(c) Show that $E_m = 0$ only at equilibrium, and $E_m > 0$ for all other values of p and v.
(d) Show that $\frac{\mathrm{d}}{\mathrm{d}t}E_m \leq 0$ for all values p, v, and that $\frac{\mathrm{d}}{\mathrm{d}t}E_m < 0$ as long as $v \neq 0$.

From the above points we may conclude, for any initial conditions $p(0)$ and $v(0)$, that:

- The mechanical energy in the system can never increase beyond its initial value.
- For any time when $v \neq 0$ the mechanical energy decreases.

Thus, no matter the initial conditions, we know that the mechanical energy in the system will never exceed the initial energy, and thus we must remain as "near" the equilibrium as we started (since going further requires additional energy). With a little more work, we can prove that $E_m \to 0$, so that we know that $p \to 0$ and $v \to 0$ as well.

This is the gist of Lyapunov's direct method, where our objective is to find a function that serves a similar role to the mechanical energy above. Unfortunately, finding such "Lyapunov functions" is often difficult.

Problem 7.18 (*22 points*) In this problem, you will implement a simulation of the quadcopter's pitch subsystem, the estimator, and the controller, for a quadcopter restricted on a rotational rig as shown below.

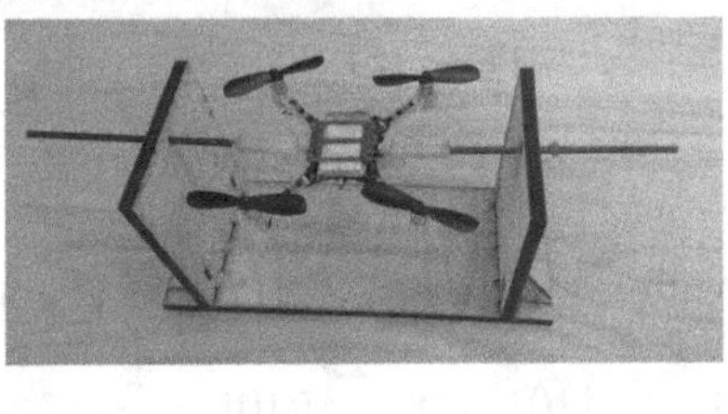

Using your favorite programming language, implement a simulation that will run for 5 s, simulating the behavior of the pitch subsystem. The system dynamics are as below, where we are using $\dot{q}_{\text{cmd}}$ as our system input:

$$\dot{\theta} = q, \qquad \dot{q} = \dot{q}_{\text{cmd}} \tag{7.78}$$

To simulate this, use a simple Euler integration, assuming the input is constant over the sampling period $\Delta t = 0.002$ s (that is, 500 Hz), where k is the time index, so that $\theta[k]$ is the value of θ at time $k\Delta t$ (note that we're using square braces, for the shorthand $\theta[k]$ meaning $\theta(t_k)$).

$$\theta[k+1] = \theta[k] + q[k]\,\Delta t + \frac{1}{2}\dot{q}_{\text{cmd}}[k]\,\Delta t^2 \tag{7.79}$$

$$q[k+1] = q[k] + \dot{q}_{\text{cmd}}[k]\,\Delta t \tag{7.80}$$

The commanded input is computed with:

$$q_{\text{cmd}}[k] := -\frac{1}{\tau_\theta}\left(\hat{\theta}[k] - \theta_{\text{cmd}}[k]\right) \tag{7.81}$$

$$\dot{q}_{\text{cmd}}[k] := -\frac{1}{\tau_q}\left(\hat{q}[k] - q_{\text{cmd}}[k]\right) \tag{7.82}$$

with $\tau_\theta = 0.12$ s and $\tau_q = 0.04$ s.

The estimator equation, as implemented is:

$$\hat{q}[k] = \gamma_2[k] \tag{7.83}$$

$$\hat{\theta}_{\text{pred}}[k+1] = \hat{\theta}[k] + \hat{q}[k]\,\Delta t \tag{7.84}$$

$$\theta_{\text{meas}}[k+1] = -\frac{\alpha_1[k]}{\left\|\underline{g}\right\|} \tag{7.85}$$

$$\hat{\theta}[k+1] = (1-\rho)\hat{\theta}_{\text{pred}}[k+1] + \rho\theta_{\text{meas}}[k+1] \tag{7.86}$$

where $\gamma_2[k]$ is the output from the rate gyroscope at time k, and $\alpha_1[k]$ is the output from the accelerometer. We choose $\rho = 0.01$ for the estimator.

You will simulate the sensors using:

$$\gamma_2[k] = q[k] \tag{7.87}$$

$$\alpha_1[k] = -\left\|\underline{g}\right\|\sin(\theta[k]) \tag{7.88}$$

Start the simulation with:

$$\hat{\theta}[0] = 0, \qquad \hat{q}[0] = 0, \qquad \theta[0] = \frac{\pi}{3}, \qquad q[0] = 0 \tag{7.89}$$

Use $\theta_{\text{cmd}}[k] = 0$ for all k.

Make a figure, with three sub-plots against time, showing

a. the true angle θ and the estimate thereof $\hat{\theta}$,
b. the angular velocity q,
c. the commanded angular acceleration $\dot{q}$.

Also provide the full listing of your code. The code must be clearly written, and well documented with comments.

Problem 7.19 (*6 points*) In this problem, we will consider the height control of a quadcopter. Let s_3 be the vehicle height, and v_3 its vertical velocity, so that $\dot{s}_3 = v_3$, and $\dot{v}_3 = u$, where u is the vertical acceleration, the input to the system

You implement a control law of the below form, whose objective is to control the vehicle to zero height and zero velocity:

$$u = -k_1 s_3 - k_2 v_3$$

You are considering three alternative controllers:

a. $k_1 = k_2 = 4$
b. $k_1 = 16, k_2 = 1$
c. $k_1 = 1, k_2 = 16$

For each of the controllers 1–3, assign which of the below responses (a)–(d) matches the choice of gain, and provide a *brief* motivation. The system is initialized at $v_3(0) = 0$, $s_3(0) = 1$, and the setpoint is always to zero height.

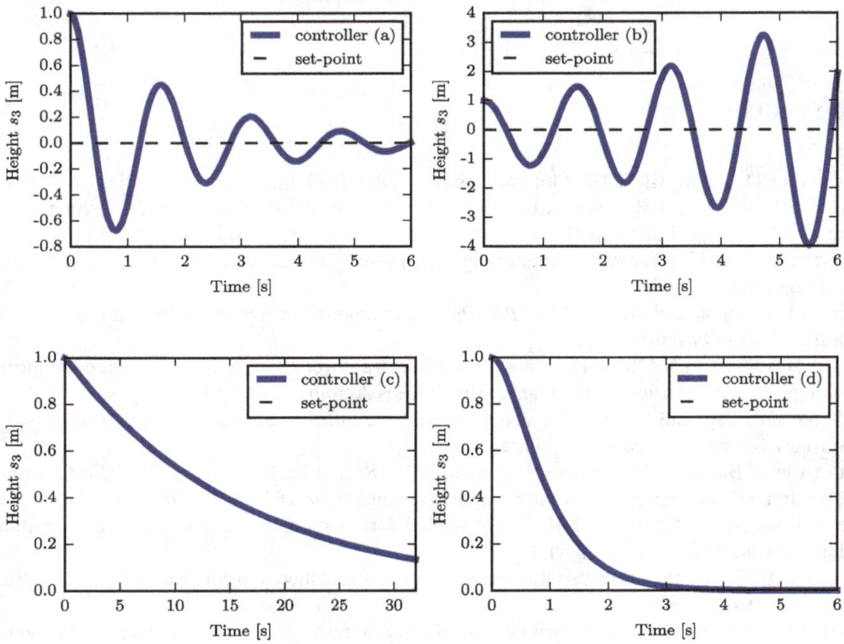

Problem 7.20 (*6 points*) You are given the below proposed control law for a quadcopter's horizontal motion along the $\underline{1}^E$ earth-fixed axis, relating the desired moment n_2 as control input u to the state $x = (s_1, v_1, \theta, q)$ and a desired position $s_{1,\text{ref}}$.

The control law is:

$$n_2 = \begin{cases} -k_1 \Delta s_1 - k_2 v_2 - k_3 \theta - k_4 & \text{if} \quad |\theta| \leq \pi / 3 \\ -k_3 \theta - k_4 q & \text{else.} \end{cases}$$

with $\Delta s_1 = s_1 - s_{1,\text{ref}}$.

Complete the below function, using C++ pseudo-code, that implements this controller, wherein: `pos1` $= s_1$, `vel1` $= v_1$, `theta` $= \theta$, `rateq` $= q$, and `posref1` $= s_{1,\text{ref}}$ (all in SI units).

You may assume that the gains are available as global variables, with the names `k1` representing k_1, and so forth for k_2 through k_4.

```
//Defined elsewhere:
//float k1, k2, k3, k4;

float controller(float posref1, float pos1, float vel1,
                 float theta, float rateq) {
    float outputTorque = 0.0f;
    // COMPLETE ME!

    return outputTorque;
}
```

References

Anderson BD, Moore JB (2007) Optimal control: linear quadratic methods. Courier Corporation

Åström KJ, Wittenmark B (1994) Adaptive control. Addison-Wesley Longman Publishing Co., Inc

Bernstein DS (2009) Matrix mathematics: theory, facts, and formulas. Princeton University Press

Bertsekas DP (2005) Dynamic programming and optimal control, vol 1, 3rd edn. Athena Scientific Belmont, MA

Borrelli F, Bemporad A, Morari M (2017) Predictive control for linear and hybrid systems. Cambridge University Press

Cao C, Hovakimyan N (2008) Design and analysis of a novel $\mathcal{L}_1$ adaptive control are architecture with guaranteed transient performance. IEEE Trans Autom Control 53(2):586591

Dullerud GE, Paganini F (2013) A course in robust control theory: a convex approach, vol 36. Springer Science & Business Media

Kaufmann E, Bauersfeld L, Loquercio A, Müller M, Koltun V, Scaramuzza D (2023) Champion-level drone racing using deep reinforcement learning. Nature 620(7976):982987

Lee T, Leok M, McClamroch NH (2017) Global formulations of Lagrangian and Hamiltonian dynamics on manifolds. Springer 13:31

Liberzon D (2011) Calculus of variations and optimal control theory: a concise introduction. Princeton University Press

Mueller MW (2018) Multicopter attitude control for recovery from large disturbances, arXiv preprint arXiv:1802.09143

Chapter 8
Advanced Topics

Sometimes the system needs an attitude adjustment.

The purpose of this chapter is to dive into some material that are not strictly necessary for understanding the preceding chapters, but which are nonetheless important or simply interesting. Specifically, we will spend a fair amount of time developing the theory for the Euler symmetric parameters (also called the quaternion of rotation), and on the way introduce the rotation vector (an intuitive, first-order tensor representation of the rotation). We then briefly derive an alternative relationship of propeller power to thrust, which allows us to reason about vehicle sizing especially, and then briefly discuss other state estimation approaches. Lastly, we look at the dynamics of unforced rotations.

8.1 Other Attitude Parametrizations

There are many attitude parametrizations, having various advantages/disadvantages. It can be shown that there is no "all-round good" parametrization (Stuelpnagel 1964), that is, a parametrization that is both continuous (no jumps) and minimal (three components for three degrees of freedom). A very complete and detailed discussion of parametrizations is given in Shuster (1993).

Recall that, in Chap. 2 we discussed the rotation matrix, and the Euler angles. The relative orientation of two frames (say, A and B) is described using the rotation tensor $\underline{\underline{R^{AB}}}$. If we coordinate the rotation tensor, we get the transformation matrix: $[T]^{AB} = \left[\underline{\underline{R^{AB}}}\right]^{A}$, a 3×3 matrix. This matrix is immediately useful, e.g., we can use it to express a tensor in a different coordinate system—consider a first-order tensor $\underline{x}$, then $\left[\underline{x}\right]^{A} = [T]^{AB}\left[\underline{x}\right]^{B}$.

M. W. Mueller, *Dynamics and Control of Autonomous Flight*, Mathematical Engineering,
https://doi.org/10.1007/978-3-031-91871-1_8

However, the matrix is cumbersome, and often unintuitive (imagine looking at a graph, with nine lines, and trying to imagine the motion of a system). An alternative is to use Euler angles; in our case typically the yaw-pitch-roll sequence. These three numbers more compactly, and more intuitively, describe the relative orientation. However, they are not perfect—computing the transformation matrix from them is complex, and error-prone; smooth motions may lead to discontinuous Euler angles, etc. We thus wish to find a better alternative—our objective in this section is to derive (and understand) the Euler symmetric parameters, commonly called 'quaternion'. To understand this, we need to first understand rotation vectors.

8.1.1 Rotation Vectors and Axis-Angle Representation

By 'Euler's theorem', the relative orientation of two frames A and B can be described as a rotation angle ρ^{BA} about a unit-length axis $\underline{n^{BA}}$. To visualize this, imagine taking an object (say, your phone), and stick an imaginary spear through it (with the direction of this spear being $\underline{n^{BA}}$). Now, rotate the spear (and thus your phone) by the angle ρ^{BA}—this now generates the new frame, B, relative to the original frame A.

We define the *rotation vector* as

$$\underline{\lambda^{BA}} := \rho^{BA}\,\underline{n^{BA}}. \tag{8.1}$$

The rotation vector has units of angle, that is rad. Thus, the length of the rotation vector is the angle between the frames, and the direction is the axis of rotation.

These are typical tensor quantities, and we can coordinate them in a coordinate system, say C, to get a list of three numbers:

$$\left[\underline{\lambda^{BA}}\right]^C = \rho^{BA}\left[\underline{n^{BA}}\right]^C \in \mathbb{R}^3 \tag{8.2}$$

Note that this is not a unique representation, we can always add 2π to the angle and keep the relationship between A and B unchanged.

From Rotation Vector to Rotation Matrix

Consider two frames, A and B; with relative orientation described by the rotation vector $\underline{\lambda^{AB}}$. Given this rotation vector, we wish to compute the transformation matrix $[T]^{AB}$. It turns out that the transformation matrix is computes as the matrix exponential of the skew-symmetric form of the coordinated rotation vector, or precisely

$$[T]^{AB} = \exp\left(\left[\underline{\underline{K}}\left(-\underline{\lambda^{AB}}\right)\right]^B\right) \tag{8.3}$$

We derive this by imagining a smooth operation, rotating frame B into frame A. Consider an intermediate frame C, and an imagined time t. At time $t = 0$, the frame C coincides with B, that is, $C = B$. We then apply a constant angular velocity to frame C, with $\underline{\omega^{CB}} = \underline{\lambda^{AB}}$ over one second. After this one (imagined) second, we have that the intermediate frame C aligns with A, i.e. at $t = 1$ we have $C = A$.

We now look at the transformation matrix between frames C and B as a function of time, $\left[\underline{T(t)}\right]^{CB}$. The initial condition specifies that $\left[\underline{T(0)}\right]^{CB} = [T]^{BB} = \left[\underline{I}\right]$. Moreover, the differential equation for the transformation matrix (taken from (3.20)) gives:

$$\left[\frac{\mathrm{d}}{\mathrm{d}t}T\right]^{CB} = [T]^{CB}\left[\underline{K}\left(\underline{\omega^{BC}}\right)\right]^{B} \tag{8.4}$$

We note that $\underline{\omega^{BC}} = -\underline{\omega^{CB}}$ (see Problem 3.3), and by our construction this is constant at $\underline{\omega^{CB}} = \underline{\lambda^{AB}}$, so that

$$\left[\frac{\mathrm{d}}{\mathrm{d}t}T\right]^{CB} = [T]^{CB}\left[\underline{K}\left(-\underline{\lambda^{AB}}\right)\right]^{B} \tag{8.5}$$

The above is a linear, time-invariant ordinary differential equation (ODE) in the matrix variable $[T]^{CB}$, where specifically $\left[\underline{K}\left(-\underline{\lambda^{AB}}\right)\right]^{B}$ is a constant matrix. The solution to this ODE is

$$\left[\underline{T(t)}\right]^{CB} = \left[\underline{T(0)}\right]^{CB}\exp\left(\left[\underline{K}\left(-\underline{\lambda^{AB}}\right)\right]^{B} t\right) \tag{8.6}$$

with $\exp(\cdot)$ the matrix exponential, defined as

$$\exp(M) := \sum_{i=0}^{\infty}\frac{M^i}{i!} \tag{8.7}$$

Convince yourself that (8.6) does, in fact, solve the ODE by substituting the infinite series.

The matrix exponential is, generally, challenging to compute, and importantly the exponential of a matrix is not the same as a matrix of the exponentials of its elements, i.e.

$$\exp\left(\begin{bmatrix} a & b \\ c & d \end{bmatrix}\right) \neq \begin{bmatrix} \exp(a) & \exp(b) \\ \exp(c) & \exp(d) \end{bmatrix} \tag{8.8}$$

Applying our boundary conditions to the solution (i.e., at $t = 0$ that $C = B$, and at $t = 1$ that $C = A$) we have

$$[T]^{AB} = \left[\underline{T(1)}\right]^{CB} \tag{8.9a}$$

$$= \left[\underline{T(0)}\right]^{CB} \exp\left(\left[\underline{K}\left(-\underline{\lambda^{AB}}\right)\right]^{B}\right) \tag{8.9b}$$

$$= [T]^{BB} \exp\left(\left[\underline{K}\left(-\underline{\lambda^{AB}}\right)\right]^{B}\right) \tag{8.9c}$$

$$= \exp\left(\left[\underline{K}\left(-\underline{\lambda^{AB}}\right)\right]^{B}\right) \tag{8.9d}$$

As an aside—note that $\left[\underline{\lambda^{BA}}\right]^{B} = \left[\underline{\lambda^{BA}}\right]^{A}$, the proof of which is left for Problem 8.1.

This form is useful for analysis, e.g., when dealing with small rotations. In that case, we may truncate the infinite series: if $\| \underline{\lambda^{AB}} \| \ll 1$, then

$$[T]^{AB} = \exp\left(\left[\underline{K}\left(-\underline{\lambda^{AB}}\right)\right]^{B}\right) = \sum_{i=0}^{\infty} \frac{\left(\left[\underline{K}\left(-\underline{\lambda^{AB}}\right)\right]^{B}\right)^{i}}{i!} \tag{8.10a}$$

$$\approx \left[\underline{I}\right] + \left[\underline{K}\left(-\underline{\lambda^{AB}}\right)\right]^{B} \tag{8.10b}$$

or, if $\left[\underline{\lambda^{AB}}\right]^{B} = (\lambda_1, \lambda_2, \lambda_3)$ then

$$[T]^{AB} \approx \begin{bmatrix} 1 & \lambda_3 & -\lambda_2 \\ -\lambda_3 & 1 & \lambda_1 \\ \lambda_2 & -\lambda_1 & 1 \end{bmatrix}. \tag{8.11}$$

Closed-Form Solution

The matrix exponential form of (8.3) is not particularly easy to use, and it turns out that there is a simple, closed-form solution to it.

$$[T]^{BA} = \cos\rho^{BA}\left[\underline{I}\right] + \left(1 - \cos\rho^{BA}\right)\left[\underline{n^{BA}}\right]^{B} \overline{\left[\underline{n^{BA}}\right]}^{B} - \sin\rho^{BA}\left[\underline{K}\left(\underline{n^{BA}}\right)\right]^{B} \tag{8.12a}$$

$$= \cos\| \underline{\lambda^{BA}} \| \left[\underline{I}\right] + \frac{\left(1 - \cos\| \underline{\lambda^{BA}} \|\right)}{\| \underline{\lambda^{BA}} \|^{2}} \left[\underline{\lambda^{BA}}\right]^{B} \overline{\left[\underline{\lambda^{BA}}\right]}^{B} - \frac{\sin\| \underline{\lambda^{BA}} \|}{\| \underline{\lambda^{BA}} \|} \left[\underline{K}\left(\underline{\lambda^{BA}}\right)\right]^{B} \tag{8.12b}$$

The derivation of this is left for Problem 8.8.

Composition of Rotations

It is tempting to imagine that, given three frames A, B, and C that the composition of rotation vectors is equivalent to their summations. However, it is important to note that (in general)

$$\underline{\lambda^{AC}} \neq \underline{\lambda^{AB}} + \underline{\lambda^{BC}} \tag{8.13}$$

The simplest way to convince yourself of this is to recall that tensor addition commutes, but second-order tensor multiplication does not, i.e.:

$$\underline{\lambda^{AB}} + \underline{\lambda^{BC}} = \underline{\lambda^{BC}} + \underline{\lambda^{AB}} \tag{8.14}$$

$$\text{but } [T]^{AB}\,[T]^{BC} \neq [T]^{BC}\,[T]^{AB} \tag{8.15}$$

To compute $\underline{\lambda^{AC}}$ from $\underline{\lambda^{AB}}$ and $\underline{\lambda^{BC}}$ is tedious, and best done via the Euler Symmetric Parameters, discussed below.

8.1.2 Euler Symmetric Parameters (ESP)

The attitude parametrization most likely to greet you if you look inside any flight controller code, or indeed any code related to three-dimensional computation, is the Euler Symmetric Parameters (ESP), which are also called the "quaternion of rotation", or simply quaternion.[1] We will spend some time describing the ESP, as they are extremely useful when implementing numerically orientations and transformation matrices.

We will use a common notation, though it is by no means unique, and loosely follow the notational convention of Zipfel (2007). What follows is best described in Shuster (1993), but note that we use a differing convention & notation. Use extreme caution when using equations from multiple sources, to ensure that they share consistent definitions and conventions (you will find, especially, often the order of the four elements swapped).

The ESP is a list of four real numbers: Let B be a frame derived by rotating frame E by the angle ρ^{BE} about a unit vector $\underline{n^{BE}}$, with $\left[\underline{n^{BE}}\right]^E = (n_1, n_2, n_3)$. We can encode this using the ESP, coordinated in system E, as

$$\{q\}^{BE} = \begin{bmatrix} q_0 \\ q_1 \\ q_2 \\ q_3 \end{bmatrix} = \begin{bmatrix} \cos\left(\frac{1}{2}\rho^{BE}\right) \\ n_1 \sin\left(\frac{1}{2}\rho^{BE}\right) \\ n_2 \sin\left(\frac{1}{2}\rho^{BE}\right) \\ n_3 \sin\left(\frac{1}{2}\rho^{BE}\right) \end{bmatrix} = \begin{bmatrix} \cos\left(\frac{1}{2}\rho^{BE}\right) \\ \left[\underline{n^{BE}}\right]^E \sin\left(\frac{1}{2}\rho^{BE}\right) \end{bmatrix} \tag{8.16}$$

[1] This is unfortunate, since the parametrization predates the quaternion formalism, and the whole concept of 'quaternion' with its hyper-complex number system is best abandoned as a historical curiosity.

Note that we use the curly braces, because this quantity doesn't fit neatly within our established tensor framework. A first property to note about the ESP is that they have norm one:

$$\left\| \{q\}^{BE} \right\|^2 = \left(\cos\left(\tfrac{1}{2}\rho^{BE}\right)\right)^2 + \left(\sin\left(\tfrac{1}{2}\rho^{BE}\right)\right)^2 \left\| \underline{n}^{BE} \right\|^2 \tag{8.17a}$$

$$= \left(\cos\left(\tfrac{1}{2}\rho^{BE}\right)\right)^2 + \left(\sin\left(\tfrac{1}{2}\rho^{BE}\right)\right)^2 = 1 \tag{8.17b}$$

Thus, we can imagine the ESP to reside on the hypersphere in $\mathbb{R}^4$.

ESP to Transformation Matrix

The transformation matrix is related to the ESP as follows from $\{q\}^{BE}$:

$$[T]^{BE} = q_0^2\left[\underline{I}\right] - \overline{\left[\underline{q_v}\right]}\left[\underline{q_v}\right]\left[\underline{I}\right] + 2\left[\underline{q_v}\right]\overline{\left[\underline{q_v}\right]} - 2q_0\left[\underline{K}\left(\underline{q_v}\right)\right] \tag{8.18a}$$

$$= \begin{bmatrix} q_0^2 + q_1^2 - q_2^2 - q_3^2 & 2\,(q_1q_2 + q_0q_3) & 2\,(q_1q_3 - q_0q_2) \\ 2\,(q_1q_2 - q_0q_3) & q_0^2 - q_1^2 + q_2^2 - q_3^2 & 2\,(q_2q_3 + q_0q_1) \\ 2\,(q_1q_3 + q_0q_2) & 2\,(q_2q_3 - q_0q_1) & q_0^2 - q_1^2 - q_2^2 + q_3^2 \end{bmatrix} \tag{8.18b}$$

We note that this requires no trigonometric functions, and is simply quadratic in the elements q_i. The computation of this is directly from (8.12a), and applying the half-angle trigonometric identities.

Because all the entries in (8.18b) appear quadratically, we note the first potential problem with the ESP: they are not unique, with specifically there being two ESP corresponding to every particular transformation matrix. The two sets of ESP are simply with all the signs flipped, so that if $\{q\}^{BA} = -\{q\}^{CA}$ then the frames B and C coincide (since $[T]^{BA} = [T]^{CA}$), even though clearly $\{q\}^{BA} \neq \{q\}^{CA}$. Thus, one must be very careful when comparing ESPs to conclude whether frames are similar or not.

Composition of Rotations

We note that, given a transformation matrix $[T]^{AB}$, we can calculate the corresponding ESP from (8.18b) by looking at the symmetric and asymmetric parts of the transformation matrix (and taking care of the two equaly valid solutions related through the sign flip). Thus, given frames A, B, and C, one can calculate $\{q\}^{CA}$ from $\{q\}^{CB}$ and $\{q\}^{BA}$ through the equation $[T]^{CA} = [T]^{CB}[T]^{BA}$. Fortunately, it turns out that this composition can be accomplished much more simply, without computing any transformation matrices. In fact, the composition of ESP is easier than the composition of transformation matrices.

It can be shown that

$$\{q\}^{CA} = K(\{q^{CB}\})\,\{q\}^{BA} \tag{8.19}$$

where we are abusing notation slightly to define the function $K(\{q^{\cdot}\})$ which maps the four dimensional column-vector $\{q\} = \left(q_0, \left[\underline{q_v}\right]\right)$ to a 4×4 matrix:

$$K(\{q\}) := q_0\left[I\right] + \begin{bmatrix} 0 & -\overline{\left[\underline{q_v}\right]} \\ \left[\underline{q_v}\right] & -\left[\underline{K}\left(\underline{q_v}\right)\right] \end{bmatrix} = \begin{bmatrix} q_0 & -q_1 & -q_2 & -q_3 \\ q_1 & q_0 & q_3 & -q_2 \\ q_2 & -q_3 & q_0 & q_1 \\ q_3 & q_2 & -q_1 & q_0 \end{bmatrix} \tag{8.20}$$

where $\left[I\right]$ is the 4×4 identity. Thus, we note that the composition of ESPs is equivalent to a matrix multiplication, similar to that of transformation matrices. However, it is a multiplication of a 4×4 matrix with a 4×1 column vector, which is less work than multiplying two 3×3 matrices. Thus, the composition of ESP is computationally simpler than that of the transformation matrix.

Differential Equations

We can derive the ESP differential equation from the composition rule. Consider our usual frames B and E, with a body rotating at angular velocity $\underline{\omega^{BE}}$ from time t_0 for a duration Δt. We know from (3.21):

$$\left[\underline{T(t_0 + \Delta t)}\right]^{EB} = \left[\underline{T(t_0)}\right]^{EB} \exp\left(\left[\underline{\Omega^{BE}}\right]^B \Delta t\right) \tag{8.21}$$

Applying to our composition rule, Eq. (8.19), we have

$$\{q\ (t_0 + \Delta t)\}^{EB} = K\left(\{q\ (t_0)\}^{EB}\right) \left\{q\left(\left[\underline{\omega^{BE}}\right]^B \Delta t\right)\right\}^{BA} \tag{8.22}$$

where we apply rotation-vector interpretation of the ESP for

$$\left\{q\left(\left[\underline{\omega^{BE}}\right]^B \Delta t\right)\right\}^{BA} = \begin{bmatrix} \cos\left(\frac{1}{2}\left\|\omega^{BE}\right\|\Delta t\right) \\ \sin\left(\frac{1}{2}\left\|\omega^{BE}\right\|\Delta t\right)\frac{1}{\left\|\omega^{BE}\right\|}\left[\underline{\omega^{BE}}\right]^B \end{bmatrix} \tag{8.23a}$$

$$\rightarrow \begin{bmatrix} 1 \\ \frac{1}{2}\Delta t\left[\underline{\omega^{BE}}\right]^B \end{bmatrix} \text{ as } \Delta t \rightarrow 0 \tag{8.23b}$$

We can then solve for the differential equation as

$$\left\{\frac{\mathrm{d}}{\mathrm{d}t}q\right\}^{EB} = \lim_{\Delta t \rightarrow 0} \frac{\{q\ (t_0 + \Delta t)\}^{EB} - \{q\ (t_0)\}^{EB}}{\Delta t} \tag{8.24a}$$

$$= \frac{1}{2}K\left(\{q\}^{EB}\right)\begin{bmatrix} 0 \\ \left[\underline{\omega^{BE}}\right]^B \end{bmatrix} \tag{8.24b}$$

This can be rewritten into a form that you may find more readily in other references, with $\left[\underline{\omega^{BE}}\right]^B = (p, q, r)$:

$$\left\{\frac{\mathrm{d}}{\mathrm{d}t}q\right\}^{EB} = \frac{\mathrm{d}}{\mathrm{d}t}\begin{bmatrix} q_0 \\ q_1 \\ q_2 \\ q_3 \end{bmatrix} = \frac{1}{2}\begin{bmatrix} 0 & -p & -q & -r \\ p & 0 & -r & q \\ q & r & 0 & -p \\ r & -q & p & 0 \end{bmatrix}\begin{bmatrix} q_0 \\ q_1 \\ q_2 \\ q_3 \end{bmatrix} \tag{8.25}$$

We note that this ODE is particularly nice. Like that of the transformation matrix, it is simply a matrix multiplication with the angular velocity, but here there are only four terms to consider, unlike the nine terms of the transformation matrix. The numerical properties of this differential equation are investigated in Problem 8.7.

8.1.3 Comparison of Attitude Representations

We have seen four different methods for encoding the relative orientation of two frames:

- the transformation matrix $[T]^{AB}$;
- the axis-angle representation, where we rotate by an angle ρ^{BE} about a unit vector $\underline{n}^{BE}$
- the yaw-pitch-roll Euler angles ψ, θ, and ϕ; and
- the Euler Symmetric parameters q_i, $i \in \{0, 1, 2, 3\}$.

Each of these has their place, and in this book we will be dealing with three specifically. In short:

- the transformation matrix $[T]^{AB}$

  ✔ useful for analysis, central to our coordinate-free derivations

  ✘ requires nine parameters to encode a 3DOF quantity, and has six constraints

  ⇒ use for analysis

- the yaw-pitch-roll Euler angles ψ, θ, and ϕ

  ✔ very intuitive, easy to visualize, maps 3DOF to three parameters

  ✘ discontinuous, numerically sensitive, differential equations are ugly (and possess a singularity)

  ⇒ use for visualization, analysis of simple (linearized) problems

- the Euler Symmetric parameters q_i, $i \in \{0, 1, 2, 3\}$

  ✔ numerically stable, easy to implement, "nice" map to transformation matrix and differential equations

  ✘ unintuitive, requires four scalars for a 3DOF object with one constraint (sum of squares of elements equals one)

  ⇒ use for implementation (coding).

This naturally leads to a question: Could one come up with another attitude representation, that combines the best of all worlds? Unfortunately, the very simple answer is "no"—for a fuller discussion see Stuelpnagel (1964).

8.2 Actuator Disk Model of Propeller Thrust and Power

We now revisit the problem of modelling a propeller, that we first touched on using a blade element model in Sect. 1.1.3. Now, instead, we neglect all the rotational aspects of the propeller (surprisingly, we're still able to derive insights!). Instead of rotating blades, we imagine a propeller as an idealized, infinitesimally thin disk, that adds energy to a fluid flowing through it, as visualized in Fig. 8.1. This energy accelerates the fluid, causing a reacting force on the disk (the propeller thrust). The end result has no direct prediction about torques (as our model does not include anything rotating!), but does allow for predictions about how power required changes as a function of the propeller size.

Assuming a non-translating propeller, we have that the aerodynamic power required by the propeller, P, is related to the thrust force f_T, the propeller radius r_p, and the density of the fluid ρ as

$$P = \frac{f_T^{\frac{3}{2}}}{r_p\sqrt{2\pi\rho}} \tag{8.26}$$

We note that the relationship between power and thrust matches our blade element model, Eq. (1.16). However, we can now also predict how changing the propeller size, r_p, will affect system power consumption.

8.2.1 Bernoulli's Law

Before we can derive (8.26), we require Bernoulli's law: Along a streamline in an inviscid, incompressible, steady flow, where there is no energy added, the static pressure p_i and speed v_i at two locations $i \in \{1, 2\}$ are related as

$$\frac{1}{2}\rho v_1^2 + p_1 = \frac{1}{2}\rho v_2^2 + p_2 \tag{8.27}$$

This can be derived from conservation of energy, though we will not do it here. Note that we assume that the flow is *inviscid*, i.e., that $Re \to \infty$; we also assume that the flow is incompressible, or $M \ll 1$.

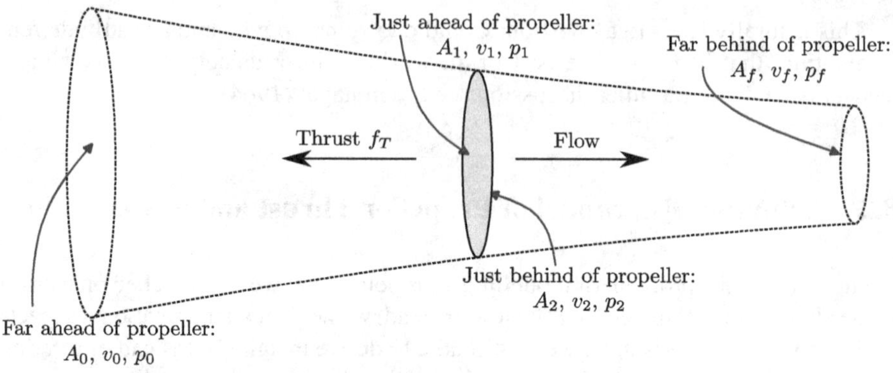

Fig. 8.1 The actuator disk model of a stationary propeller, which accelerates air through the 'actuator disk' (shown in grey) to produce a thrust force f_T

8.2.2 *Derivation*

Referring to Fig. 8.1: we will analyze the flow of an infinitesimal packet of air moving from far ahead of the propeller to far behind it. We identify 4 surfaces at which we will analyze the flow. Because the flow is assumed to be incompressible, the density is constant at ρ. At each surface i, we consider the area A_i, the speed of the flow v_i, and the static pressure p_i.

0: This is far ahead of the actuator disk, where the air is at rest and at ambient conditions. We assume at this surface:

- The air is at rest, $v_0 = 0$.
- The static pressure equals the ambient pressure, $p_0 = p_\infty$.

1: This is just ahead of the actuator disk.

- The size of this disk is the size of the propeller, $A_1 = A_p = \pi r_p^2$ with r_p the propeller radius.
- The flow speed at this point is typically called the *induced velocity*, $v_1 = v_i$.

2: This is just behind the actuator disk, that is, just after the propeller has acted on the air. Because our propeller is an infinitesimally thin disk, surfaces 1 and 2 are infinitesimally close. We assume:

- The size is the same as surface 1, $A_2 = A_1 = A_p$.
- Through the disk, the propeller adds power P to the flow while producing thrust f_T.

f: This is far behind the actuator disk, far enough that the static pressure has equalized with the ambient pressure:

- Thus $p_f = p_\infty$.

First, we note that the flow speed is constant through the propeller disk. We show this by considering the conservation of mass along the flow, and equating *mass flux* at surface 1 and 2:

$$\rho v_1 A_1 = \rho v_2 A_2 \tag{8.28}$$

Since $A_1 = A_2 = A_p$, we have that the speed at surface 2 also equals the induced velocity: $v_2 = v_1 = v_i$.

Between surface 0 and 1 we can apply Bernoulli's law, and then substitute our conditions at surface 0:

$$\frac{1}{2}\rho v_0^2 + p_0 = \frac{1}{2}\rho v_1^2 + p_1 \tag{8.29a}$$

$$p_\infty = \frac{1}{2}\rho v_i^2 + p_1 \tag{8.29b}$$

Similarly, between surface 2 and f:

$$\frac{1}{2}\rho v_2^2 + p_2 = \frac{1}{2}\rho v_f^2 + p_f \tag{8.30}$$

$$\frac{1}{2}\rho v_i^2 + p_2 = \frac{1}{2}\rho v_f^2 + p_\infty \tag{8.31}$$

Taking the difference between (8.29b) and (8.31), we can solve the downstream velocity v_f from the pressure difference $p_2 - p_1$ as

$$v_f = \sqrt{\frac{2\,(p_2 - p_1)}{\rho}} \tag{8.32}$$

The pressure difference at the actuator disk, meanwhile, creates a thrust force f_T:

$$f_T = A_2 p_2 - A_1 p_1 = A_p\,(p_2 - p_1) \tag{8.33}$$

At the actuator disk, moreover, the propeller adds power P to the flow. This added power changes the energy flux through the flow, related to the mass flow rate $\dot{m}$:

$$P = \frac{1}{2}\dot{m}v_f^2 - \frac{1}{2}\dot{m}v_0^2 \tag{8.34}$$

Note that this neglects potential energy. Substituting $v_0 = 0$, Eq. (8.32) for v_f, and computing the mass flow rate as $\dot{m} = \rho A_p v_i$, and substituting the thrust value from (8.33), we get:

$$P = v_i\, f_T \tag{8.35}$$

By Newton's law, the thrust will also change the momentum of the flow. We can compare the momentum flux at surfaces 0 and f (going from $v_0 = 0$ to v_f), and note that the only external force on the flow is the thrust f_T to get:

$$f_T = \dot{m}v_f - \dot{m}v_0 = \dot{m}A_p v_i v_f \tag{8.36}$$

Substituting now (8.32) and then (8.35) we get the desired result:

$$P = \frac{f_T^{\frac{3}{2}}}{r_p\sqrt{2\pi\rho}}. \tag{8.37}$$

8.3 Principled Model-Based State Estimation

In Chap. 5 a series of intuitive state estimators were presented which, eventually, would be sufficient to have a quadcopter hover. That estimator is intuitive, should be fairly easy to interpret, and requires extremely little computational power. However, it was created *ad hoc*, and relied on many questionable assumptions. For example, it would be difficult to extend to additional sensors, different capabilities, etc.

For serious applications, a different approach is required. Especially in aerospace application, this typically takes the form of a *model-based* state estimator, typically basing off the Kalman filter. The details of this are vastly outside the scope of these notes; but below we provide a brief sketch. This is presented as a formulaic set of steps; in reality a successful application requires significant domain expertise and adaptation to a particular problem.

At its heart, a Kalman filter is a two-step estimator, like that which we derived in Chap. 5: a *prediction step* which moves our estimate forward through time, and then an *update step* which incorporates measurements to update our estimate. The actual work in implementing a Kalman filter lies in creating a model (of a very specific form) of the system we are interested in. After the model is created, the actual implementation of a filter is (typically) a purely rote exercise of coding the equations.

Abstractly, we are first required to create a model of our system that has the following form:

$$x(k) = q\left(k, x(k-1), u(k-1), v(k-1)\right) \tag{8.38}$$
$$z(k) = h\left(k, x(k), w(k)\right) \tag{8.39}$$

wherein:

- The system is modelled in discrete time, with k being the *time step*. Of course, our systems are naturally continuous time (due to Newton's law being expressed as a differential equation)—this is not a big deal, and we must simply implement an appropriate time-discretization scheme (as we also had to do in Chap. 5).

- The system state at time k is the vector $x(k)$. The state is often the same as the state one would use for control design, but does not need to be identical.
- The function $q(\cdot)$ is the *dynamics model* of the system, and describes how the system's future develops:
  - The function may be time-varying, hence the dependence on the time k.
  - The present state obviously depends on the past state of the system, $x(k-1)$.
  - The known input to the system is the vector $u(k-1)$, e.g. the motor forces commanded at the last time step.
  - The modelling error, or prediction uncertainty, is encoded through the *process noise vector* $v(k-1)$. This is a vector-valued random vector, which is assumed to be white (i.e., uncorrelated over time) and zero-mean. The process noise captures the fact that our predictions are imprecise, and even if we knew the past state of the system exactly, there would be uncertainty in its present state (e.g., due to wind forces).
- The function $h(\cdot)$ is the *measurement model*:
  - Its output is the measurement vector, $z(k)$, a vector containing the output of all sensors at time k.
  - The measurement function may be time-varying, most commonly because some sensors are only available intermittently, or operate at different frequencies (e.g., an IMU running 500 Hz vs. a GPS running 10 Hz).
  - Of course, the measurements are related to the state of the system, $x(k)$.
  - Lastly, the uncertainty in the measurement model is captured by the *measurement noise*, $w(k)$. This is another vector-valued random vector, also assumed to be white; it encodes all sources of errors in our sensors. Such errors could include what we typically think of as noise (e.g., thermal noise on an IMU), but may also include other modelling errors (such as linearization errors).

The job of the state estimation engineer is thus to take the complex physical system, and reduce it to a stochastic abstraction of the form (8.38) and (8.39). This is a more challenging task than it may appear at first glance, especially creating a consistent, reasonably accurate but also reasonably compact probabilistic description of the random variables (including the initial state uncertainty).

Once a model is finalized, a fairly straight-forward translation to a Kalman filter is possible, which eventually results in a state estimator of the form:

$$\hat{x}_p(k) = q\left(k, \hat{x}_m(k-1), u(k-1), 0\right) \tag{8.40}$$

$$\hat{z}(k) = h\left(k, \hat{x}_p(k), 0\right) \tag{8.41}$$

$$\hat{x}_m(k) = \hat{x}_p(k) + K(k)\left(z(k) - \hat{z}(k)\right) \tag{8.42}$$

wherein:

- The vector $\hat{x}_m(k)$ is our estimate of the state $x(k)$, having incorporated all measurements up to time k.

- The vector $\hat{x}_p(k)$ is the *one step prediction*, i.e., our best guess of the state at time k, but *before* we get the latest measurement $z(k)$. Thus, Eq. (8.40) is the prediction step of our estimator. Note that we use the mean value of 0 for the noise.
- We predict the measurement as $\hat{z}(k)$, i.e., this is our best guess of what the sensor measurement will be at time k.
- Lastly, our state update (or correction step) happens in (8.42). This includes the *Kalman gain* $K(k)$, a matrix whose computation is too complicated to repeat here (but still formulaic). Notice that we correct our state estimate proportionally to the difference between our true measurement $z(k)$ and our predicted measurement $\hat{z}(k)$.

Interestingly, the Kalman filter generates a *confidence* value in addition to the state estimate $\hat{x}_m(k)$. This is expressed as a variance, and quantifies our belief in the estimation error size (again, though, the details are beyond the scope of this course).

There are peculiarities when doing Kalman filters over complex systems, relevantly here when doing Kalman filters with a state that includes a rotation. For an easy-to-apply approach to applying Kalman filter with attitude states, see Mueller et al. (2017).

Thus, hopefully, the advantage of these approaches is clear. If our system has an additional sensor, this translates into adapting the function h. Moreover, we need not create "pseudo-measurements" of our states, as we did in Chap. 5; instead we simply need to model the dynamics and the sensors. The estimator will extract the present state information in every sensor.

A note about language: where the system dynamics and measurement equations are not linear (as they would be for the systems we deal with), one would use the *extended Kalman filter (EKF)* or *unscented filter (UF)*. Strictly, the Kalman filter was originally defined for linear systems.

A good general overview of model-based state estimation is the book by Simon (2006); while the book by Bar-Shalom et al. (2001) provides a great overview of estimation focusing on tracking applications.

8.4 The Dynamics of Unforced Rotations

Here is a fun experiment: take a rectangular prism (perhaps your smartphone), and identify the three rectangular faces. Note these as the largest, mid-sized, and smallest (largest obviously where the screen is, etc.). Imagine then an axis at the center of each face, aligned with each face's normal direction, as illustrated in Fig. 8.2. Now, carefully (and ideally over a padded surface), try tossing your phone so that it has a clean rotation about one of these axes, and try this for each of the three.[2] Most likely, with very little effort, you can get a nice rotation about the largest face, and the smallest face, but you cannot get a nice rotation about the middle face—the phone

[2] Best do this with someone else's phone, in case you drop & break it.

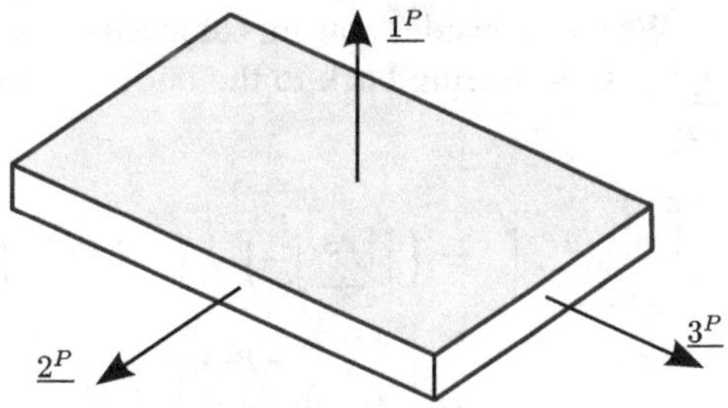

Fig. 8.2 The three principal axes of inertia of a rectangular prism, such as a smartphone

seems to always rotate in a more complex way than you intended, no matter how cleanly you try to launch it. In this subsection we will see that explaining this is a neat combination of our dynamics and stability tools.

Understanding these dynamics not only makes for a great party trick that gives you an excuse to wheel out a whiteboard during thanksgiving dinner, but is also key to some interesting flight mechanics, such as allowing a multicopter to continue controlled flight after the failure of an actuator (Mueller and D'Andrea 2015), or even creating controllable flying systems with only one propeller (and no other actuators or aerodynamic surfaces) (Zhang et al. 2019).

We recall from our analysis of the inertia tensor in Sect. 4.2.1, that every rigid body has a body-fixed frame so that the inertia tensor in that frame is diagonal (with the directions of this frame being the 'principal axes of inertia'). For objects with mirror symmetry (such as our phone), these are easy to identify (the axes align with the normal of the plane of mirror symmetry); and thus the face normals we were considering before are our phone's principal axes of inertia. For an asymmetric body, the axes may be harder to identify but nonetheless exist.

For the purpose of this section, let us define a rigid body's P frame as that frame which yields

$$\left[\underline{J_C^B}\right]^P = \begin{bmatrix} J_{11} & 0 & 0 \\ 0 & J_{22} & 0 \\ 0 & 0 & J_{33} \end{bmatrix} \tag{8.43}$$

with $J_{11} \geq J_{22} \geq J_{33}$, with C the body's center of mass. In other words, the directions of P are the principal axes, and we arrange them so that $\underline{1^P}$ is the largest, and $\underline{3^P}$ is the smallest moment of inertia. We will define the angular velocity of the body with respect to our (assumed inertial) earth-fixed frame E as

$$\left[\underline{\omega^{BE}}\right]^P = \begin{bmatrix} \omega_1 \\ \omega_2 \\ \omega_3 \end{bmatrix} \tag{8.44}$$

We now consider how this object moves when there are no external moments, i.e., $\underline{n_C} = 0$. Referring back to the body-fixed expression for Euler's law of (4.13), we get

$$\left[\frac{\mathrm{d}}{\mathrm{d}t}\underline{\omega^{BE}}\right]^P = \left(\left[\underline{J_C^B}\right]^P\right)^{-1}\left(-\left[\underline{\Omega^{BE}}\right]^P\left[\underline{J_C^B}\right]^C\left[\underline{\omega^{BE}}\right]^C + \left[\underline{n_C}\right]^P\right) \tag{8.45a}$$

$$= \begin{bmatrix} \frac{J_{22}-J_{33}}{J_{11}}\omega_2\omega_3 \\ \frac{J_{33}-J_{11}}{J_{22}}\omega_1\omega_3 \\ \frac{J_{11}-J_{22}}{J_{33}}\omega_1\omega_2 \end{bmatrix} \tag{8.45b}$$

The angular velocity is thus a three-state, autonomous, nonlinear dynamic system with state vector $x = (\omega_1, \omega_2, \omega_3)$, which we can write more compactly by introducing the constants $a_1 := \frac{J_{22}-J_{33}}{J_{11}}$, $a_2 = \frac{J_{33}-J_{11}}{J_{22}}$, and $a_3 := \frac{J_{11}-J_{22}}{J_{33}}$, so that

$$\dot{x} = \begin{bmatrix} \dot{\omega}_1 \\ \dot{\omega}_2 \\ \dot{\omega}_3 \end{bmatrix} = \begin{bmatrix} a_1\omega_2\omega_3 \\ a_2\omega_1\omega_3 \\ a_3\omega_1\omega_2 \end{bmatrix} = f(x) \tag{8.46}$$

We now follow the pattern of Chap. 7 to analyze this. First, we find the equilibria: any set of three angular velocities $(\omega_1^*, \omega_2^*, \omega_3^*)$ so that the right-hand side of (8.46) is zero. Trivially, if $a_1 = a_2 = a_3 = 0$, then any value ω_i^* is an equilibrium. This corresponds to rigid bodies where $J_{11} = J_{22} = J_{33}$, for example, a ball. Of course, this matches our intuition—I can toss a ball with any initial angular velocity, and it will continue rotating as it had left my hands without anything exciting happening.

Let us, then, consider asymmetric bodies, where $J_{11} \neq J_{22} \neq J_{33}$, so that $a_1 \neq 0$, $a_2 \neq 0$, and $a_3 \neq 0$. We note that there are three 'families' of non-trivial equilibria:

$$x^* = \begin{bmatrix} w_1^* \\ 0 \\ 0 \end{bmatrix} \quad \text{or} \quad x^* = \begin{bmatrix} 0 \\ w_2^* \\ 0 \end{bmatrix} \quad \text{or} \quad x^* = \begin{bmatrix} 0 \\ 0 \\ w_3^* \end{bmatrix} \tag{8.47}$$

i.e., any rotation about a principal axis is an equilibrium (but a rotation with non-zero angular velocity component along two different principal axes is not an equilibrium).

Next, we can investigate the stability of these equilibria, with our typical first-order approximation $\dot{z} = Az$, where we calculate:

$$A = \frac{\partial}{\partial x}f(x^*) = \begin{bmatrix} 0 & a_1\omega_3^* & a_1\omega_2^* \\ a_2\omega_3^* & 0 & a_2\omega_1^* \\ a_3\omega_2^* & a_3\omega_1^* & 0 \end{bmatrix} \tag{8.48}$$

For the symmetric system with equal inertia, where $a_i = 0$, we of course have $A = 0$, and all eigenvalues of A are also zero, so that the indirect method provides no insight in stability. This is not too surprising, since any angular velocity is an equilibrium.

Next, we look at the interesting cases with all $a_i \neq 0$. Because we ordered our axes so that $J_{11} > J_{22} > J_{33}$, we know that $a_1 > 0, a_2 < 0$, and $a_3 > 0$. Considering the case $x^* = (\omega_1^*, 0, 0)$, we have

$$A = \begin{bmatrix} 0 & 0 & 0 \\ 0 & 0 & a_2\omega_1^* \\ 0 & a_3\omega_1^* & 0 \end{bmatrix} \tag{8.49}$$

which has eigenvalues $\lambda \in \{0, \pm\sqrt{a_2 a_3}\,|\omega_1^*|\}$. Because $a_2 < 0$ and $a_3 > 0$, this means that all eigenvalues lie on the imaginary axis, so that the indirect method can provide no conclusion. For the case with $\omega_3^* \neq 0$, we have by the same logic $\lambda \in \{0, \pm\sqrt{a_1 a_2}\,|\omega_3^*|\}$, which are again all three purely imaginary eigenvalues.

However, for $\omega_2^* \neq 0$, the eigenvalues are $\lambda \in \{0, \pm\sqrt{a_1 a_3}\,|\omega_3^*|\}$ which are strictly real (since $a_1 a_3 > 0$), so that there is one eigenvalue with positive real part. Because we have an eigenvalue with positive real part, the indirect method allows us to conclude that this is an *unstable equilibrium*. This explains why we cannot get a nice rotation about our phone's intermediate face—there is an equilibrium motion there, but the equilibrium is unstable, and any initial imperfection will cause the system to deviate from the equilibrium. Depending on the inertia values, it may be *very* unstable, in the sense of having a large positive eigenvalue. Thus, we have our first fun fact—you can demonstrate an exponentially unstable system simply by tossing a phone (and then you can show this derivation to explain it)!

Frustratingly, for the two equilibria with $\omega_1^* \neq 0$ or $\omega_2^* \neq 0$, the indirect method provided no conclusion about stability. By tossing our phones, we have the intuition that over the duration of a typical free flight (before we have to catch the phone again) both of these rotations continue as started. We may be tempted to assume that they are both equivalent, and both stable in the sense of Lyapunov. This turns out to be false, and we will show this in a slightly roundabout way which will require the concept of *kinetic energy*.

Definition 8.1 A rigid body B with center of mass C has a total kinetic energy k^{BR} with respect to reference frame R given by

$$k^{BR} := \frac{1}{2}\overline{\underline{\omega}^{BR}}\,\underline{\underline{J}}_C^B\,\underline{\omega}^{BR} + \frac{1}{2}m^B\,\overline{\underline{v}_C^R}\,\underline{v}_C^R \tag{8.50}$$

where $\underline{\underline{J}}_C^B$ is the body's inertia tensor about its center of mass, and m^B is the body's mass.

Note that this is a scalar quantity, and results from an invariant description, i.e. the energy doesn't depend on the coordinate system used. However, the energy is defined *relative to a reference frame*.

Assume that our rigid body B initially has some amount of angular momentum, which we will call μ, with $\mu > 0$. For the sake of simplicity, assume we define our earth-fixed system so that initially

$$\underline{l_C^{BE}} = \mu \, \underline{1^E} \tag{8.51}$$

i.e., our body is rotating so that the angular momentum vector aligns with $\underline{1^E}$. Euler's law states that, in the absence of an applied moment, the angular momentum is constant with respect to an inertial frame, so that $\underline{l_C^{BE}} = \mu \, \underline{1^E}$ for all time. If we now orient the body so that it is rotating about the $\underline{1^P}$ axis so that $\underline{\omega^{BE}} = \omega_1^* \, \underline{1^P}$, we have angular momentum

$$\left\| \underline{J_C^B} \, \underline{\omega^{BE}} \right\| = \mu \tag{8.52}$$

Coordinating in the P system, we have $\omega_1^* J_{11} = \mu$. The kinetic energy that this rotation has (neglecting the translational velocity component) is then

$$k^{BE} = \frac{1}{2} \overline{\underline{\omega^{BR}}} \, \underline{J_C^B} \, \underline{\omega^{BR}} = \frac{1}{2} J_{11} \left(\omega_1^*\right)^2 \tag{8.53}$$

which we can relate to the angular momentum as $k^{BE} = \frac{1}{2}\mu^2 / J_{11}$. By a similar argument, the kinetic energy when rotating about $\underline{3^B}$ with the same angular momentum is $k^{BE} = \frac{1}{2}\mu^2 / J_{33}$.

Because $J_{11} > J_{33}$, the system has higher kinetic energy rotating about the $\underline{3^P}$ axis, than about the $\underline{1^P}$ axis. If we can thus imagine a process that can convert the system's kinetic energy into something else, we can conclude that the rotation around $\underline{3^P}$ is *unstable in the sense of Lyapunov*, as any loss of energy pushes the system away from the equilibrium. In atmospheric flight, this is typically not noticeable, as external disturbances will have bigger effects on the system than this (e.g., aerodynamic drag will typically reduce kinetic energy).

However, in space, this effect can dominate. One such mechanism by which energy can be transferred may be a liquid inside a tank of a rocket; any slight structural vibration of the rocket will be transmitted to the liquid, whose internal motion will be transformed to heat, and the heat will be radiated away from the rocket. Such a process requires no external moment, so that the angular momentum of the system remains unchanged at $\underline{l_C^{BE}} = \mu \, \underline{1^E}$. For a system rotating about $\underline{1^P}$, it is impossible to transform the kinetic energy into anything else, because any lower kinetic energy would require a change in momentum, which would require an external moment.

This surprising fact contributed to the failure of the Explorer I satellite; which was designed to spin around its long axis (the axis with smallest inertia). The satellite, instead of continuing to rotate about the long axis, settled into a rotation perpendicular to that axis. For this reason, modern rotating satellites rotate about a 'fat' axis, rather than a 'skinny' axis, see e.g. the satellites shown in Fig. 8.3. The book by Maruskin Maruskin (2012) has more information.

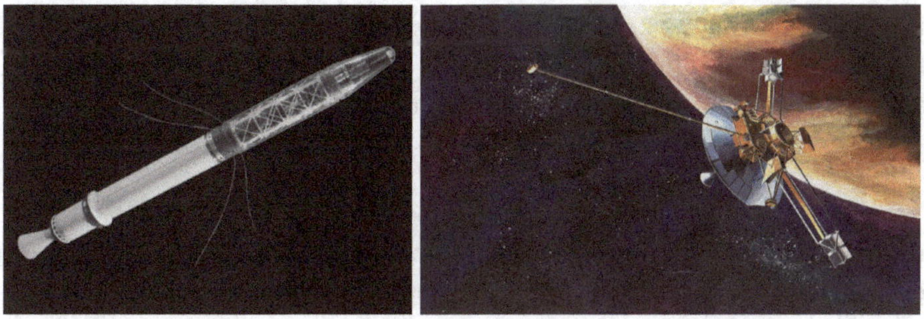

Fig. 8.3 The Explorer 1 satellite, shown on the left, was designed to rotate around its axis of smallest inertia. Pioneer 10, shown on the right, rotates about its axis of greatest inertia. Images from NASA

Problems for Chapter 8

Problem 8.1 (*4 points*) Given are two frames A and B, with associated axis-angle representation with angle ρ^{AB} and axis $\underline{n^{AB}}$. Show that the rotation axis is the same when expressed in either coordinate system, that is, show that $\left[\underline{n^{AB}}\right]^A = \left[\underline{n^{AB}}\right]^B$.

Hint: First show that, for any first order tensor $\underline{x}$ *and coordinate system* C, *it holds that* $\left[\underline{K}(\underline{x})\right]^C \left[\underline{x}\right]^C = (0, 0, 0)$.

Problem 8.2 (*4 points*) Let E and B be respectively the Earth- and body-fixed frames, and let D be a desired body frame. These frames are related through the following sets of Euler Symmetric Parameters:

$$\{q\}^{BE} = (0, 1, 0, 0) \text{ and } \{q\}^{ED} = \left(\tfrac{1}{2}\sqrt{2}, 0, \tfrac{1}{2}\sqrt{2}, 0\right)$$

Compute the angle between the B and D frames, ρ^{BD}.

Problem 8.3 (*6 points*) Use (8.18b) and the equation for the rotation matrix as a function of the Euler angles to explicitly compute the Euler angles ψ, θ and ϕ from the elements q_i of the Euler symmetric parameters.

Problem 8.4 (*8 points*) Derive relationship between Euler Symmetric Parameters and the transformation matrix (8.18b) from that between the rotation vector and the transformation matrix (8.12a).

Problem 8.5 (*8 points*) Let A and B be two frames, with $\{q\}^{AB} = \left(q_0, \left[\underline{q_v}\right]\right)$ the ESP describing their configuration. Show that the inverse rotation is given by $\{q\}^{BA} = \left(q_0, -\left[\underline{q_v}\right]\right)$.

Problem 8.6 (*8 points*) Let A, B, and C be three frames, with corresponding sets of ESPs $\{q\}^{BA} = \left(q_0^{BA}, \left[\underline{q_v^{BA}}\right]\right)$ and $\{q\}^{CB} = \left(q_0^{CB}, \left[\underline{q_v^{CB}}\right]\right)$. Show that the distance (i.e., the angle) between frames A and C can be calculated as

$$\rho^{CA} = 2 \arccos\left(q_0^{CB} q_0^{BA} - \overline{\left[\underline{q_v^{CB}}\right]}\left[\underline{q_v^{BA}}\right]\right).$$

Problem 8.7 (*25 points*) In this problem, we will be investigating the numerical integration of orientations using different representations. This problem requires substantial programming. The goal is to investigate numerical issues, but also get a feeling for how to implement different representations.

Consider a body, with attached frame B, rotating relative to an earth-fixed frame E. At time $t = 0$, the frames coincide, so that $[T]^{EB} = \left[\underline{I}\right]$. The body rotates at a constant angular velocity $\left[\underline{\omega^{BE}}\right]^B = (1, 10, 0)$ rad/s.

Using a simple forward-euler scheme, write computer code to numerically integrate this angular velocity for $t \in [0, 1]$s using a time step of $\Delta t = 0.01$s. Do this for three different ways of representing the attitude: (1) as a transformation matrix directly, (2) as Euler angles, and (3) as Euler symmetric parameters.

Implement the integration as follows, computing the rates using the equations:

- For the transformation matrix, $\left[\frac{\mathrm{d}}{\mathrm{d}t} T\right]^{EB}$ use (3.20)
- For the Euler angles, $\dot{\psi}$, $\dot{\theta}$, and $\dot{\phi}$ use (3.23)
- For the ESP, $\left\{\frac{\mathrm{d}}{\mathrm{d}t} q\right\}^{EB}$, use (8.25)

Then, integrate using a standard forward Euler discretization

$$\left[\underline{T(t+\Delta t)}\right]^{EB} = \left[\underline{T(t)}\right]^{EB} + \left[\frac{\mathrm{d}}{\mathrm{d}t} T(t)\right]^{EB} \Delta t$$

$$\phi(t+\Delta t) = \phi(t) + \left(\frac{\mathrm{d}}{\mathrm{d}t}\phi(t)\right)\Delta t$$

$$\theta(t+\Delta t) = \theta(t) + \left(\frac{\mathrm{d}}{\mathrm{d}t}\theta(t)\right)\Delta t$$

$$\psi(t+\Delta t) = \psi(t) + \left(\frac{\mathrm{d}}{\mathrm{d}t}\psi(t)\right)\Delta t$$

$$\{q\,(t+\Delta t)\}^{BE} = \{q\,(t)\}^{BE} + \left\{\frac{\mathrm{d}}{\mathrm{d}t} q\right\}^{BE} \Delta t$$

Because the angular velocity is constant throughout the motion, we can compute the exact solution too (to evaluate how well we're approximating). The exact solution corresponds to a rotation vector $\left[\underline{\lambda^{EB}(1)}\right]^B = (-1, -10, 0)$.

We will define an error metric to compare two matrices A and B as the maximum singular value of $(A - B)$, i.e.

$$\epsilon(A, B) = \sigma_{\max}(A - B)$$

Note that, if A and B are equal, then the error is zero. *Aside: Technically, this is an "induced norm" of $A - B$, induced over the Euclidean norm (the 2-norm).* You can compute this error as follows: in Python:

```
import numpy as np # in case you don't have it yet
  def epsilon(A, B):
      C = A - B # the difference
      D = C*C.T # squaring the matrix
      e = np.max(np.linalg.eigvals(D))
      return np.sqrt(e)
```

In Matlab, you can use the following routine:

```
% assume we have matrices A and B
C = A - B;
D = C*C';
epsilon = sqrt(max(eig(D))); % this is the error metric.
```

Deliverables. For your solution, you must submit:

a. *(5 pt)* The code with which you've solved the problem.
b. *(6 pt)* The final value you get for each of the three approximations you used.
c. *(6 pt)* The final value you get for each of the three approximations you used, expressed as a transformation matrix (i.e. convert Euler angles and Euler symmetric parameters to transformation matrix).
d. *(3 pt)* The exact solution, computed from the rotation vector.
e. *(5 pt)* The error in the transformation matrix for each approximation, compared to the true value, using $\epsilon(\cdot, \cdot)$ as defined above. Comment on how the three methods performed, compared to one another.
f. *Extra credit for (5 pt), optional* Explain why the error metric $\epsilon(\cdot, \cdot)$ makes sense in this context. You'll need to explain this using the definition of an induced norm, and how we use transformation matrices. Your explanation should be brief (5 sentences or so).

Some other things to try: (these will not be graded, but are ideas for things that you can try out, to get a better intuition of what's happening)

g. A valid set of Euler symmetric parameters must have norm 1. Let's see how that affects things:

    (i) What is the norm of the numerically integrated ESP that you computed before, at the end ($t = 1$)? Given that this is not exactly 1, is the matrix computed using (8.18b) a "real" transformation matrix? Compute the matrix's determinant (should be exactly $+1$), and check whether the columns are orthogonal (compute the scalar products for each combination of two columns, the answer should always be exactly zero).

(ii) Normalize the ESP with $\{q\}^{BE} \leftarrow \frac{1}{\|\{q\}^{BE}\|}\{q\}^{BE}$, so that it has length 1. Compute the corresponding transformation matrix, and compute the error ϵ with this. How does this error compare to what you had before?

(iii) Instead of only doing the normalization at the final time instant, perform the normalization of the ESP at every integration time step. What is the final error for the transformation matrix?

h. We know that the precision of a numerical integration scheme generally depends on the size of the time step, Δt. Specifically, we expect the error to grow as Δt^2 for a first-order method (as we've implemented, see https://en.wikipedia.org/wiki/Euler_method#Local_truncation_error).

(i) Compute the above errors for values of $\Delta t \in \{0.001, 0.01, 0.1, 1\}$s.

(ii) For a given integration error target (say we want $\epsilon(\cdot,\cdot) = 0.1$), what is the time step required (for the angular velocity and initial conditions provided) for each of the three representations used? Compute this for the ESP for both the cases with and without normalization.

Problem 8.8 (*12 points*) In this problem we set out to prove the closed-form solution for the transformation matrix resulting from a rotation vector of (8.12b), i.e. prove the following relationship:

$$[T]^{BA} = \exp\left(-\left[\underline{K}\left(\underline{\underline{\lambda^{BA}}}\right)\right]^B\right) \tag{8.54}$$

$$= [\underline{I}] + \left(-\left[\underline{K}\left(\underline{\underline{\lambda^{BA}}}\right)\right]^B\right) + \frac{1}{2!}\left(-\left[\underline{K}\left(\underline{\underline{\lambda^{BA}}}\right)\right]^B\right)^2 + \frac{1}{3!}\left(-\left[\underline{K}\left(\underline{\underline{\lambda^{BA}}}\right)\right]^B\right)^3 + \cdots \tag{8.55}$$

$$= \cos\|\underline{\lambda^{BA}}\|\,[\underline{I}] + \frac{(1-\cos\|\underline{\lambda^{BA}}\|)}{\|\underline{\lambda^{BA}}\|^2}\,[\underline{\lambda^{BA}}]^B\,\overline{[\underline{\lambda^{BA}}]}^B - \frac{\sin\|\underline{\lambda^{BA}}\|}{\|\underline{\lambda^{BA}}\|}\left[\underline{K}\left(\underline{\underline{\lambda^{BA}}}\right)\right]^B \tag{8.56}$$

To save time and space, we will use the short-hands $[\underline{L}] := \left[\underline{K}\left(\underline{\underline{\lambda^{BA}}}\right)\right]^B$ and $[\underline{\lambda}] := [\underline{\lambda^{BA}}]^B$, and we'll let $[\underline{\lambda^{BA}}]^B =: (\lambda_1, \lambda_2, \lambda_3)$.

We will require the infinite series expansions for sine and cosine:

$$\sin x = \sum_{n=0}^{\infty} \frac{(-1)^n x^{2n+1}}{(2n+1)!} = x - \frac{x^3}{3!} + \frac{x^5}{5!} - \frac{x^7}{7!} + \cdots \tag{8.57}$$

$$\cos x = \sum_{n=0}^{\infty} \frac{(-1)^n x^{2n}}{(2n)!} = 1 - \frac{x^2}{2!} + \frac{x^4}{4!} - \frac{x^6}{6!} + \cdots \tag{8.58}$$

We know that (8.55) follows from (8.54) by the definition of the matrix exponential. Our objective is to show that (8.55) implies (8.56). To do this:

a. Show that

$$[\underline{L}]^2 = [\underline{\lambda}]\overline{[\underline{\lambda}]} - \overline{[\underline{\lambda}]}[\underline{\lambda}][\underline{I}] \tag{8.59}$$

b. Show that

$$[\underline{L}]^3 = -\left(\overline{[\underline{\lambda}]}[\underline{\lambda}]\right)[\underline{L}] = -\|[\underline{\lambda}]\|^2[\underline{L}] \tag{8.60}$$

c. Show that 8.60 implies that, for integer n:

(i) For even powers:

$$[\underline{L}]^{2n} = -(-1)^n \|[\underline{\lambda}]\|^{2n-2}[\underline{L}]^2 \text{ for } n \geq 1 \tag{8.61}$$

(ii) For odd powers:

$$[\underline{L}]^{2n+1} = (-1)^n \|[\underline{\lambda}]\|^{2n}[\underline{L}] \text{ for } n \geq 0 \tag{8.62}$$

d. By partitioning (8.55) into even and odd powers, show that (8.56) follows therefrom.

Problem 8.9 (*6 points*) Let E, B, and S be respectively the Earth-, body-fixed, and sensor frames. These frames are related through the following rotation vectors:

- $[\underline{\lambda}^{BE}]^B = (\pi/3, 0, 0)$ rad
- $[\underline{\lambda}^{SB}]^S = (0, 0, \pi)$ rad

Using the Euler symmetric parameters, compute the rotation vector $[\underline{\lambda}^{SE}]^S$.

Problem 8.10 (*8 points*) Let A, B, and C be three coordinate systems. Solve $\{q\}^{AB}$ so that

$$\{q\}^{AC} = \begin{bmatrix} \frac{\sqrt{3}}{2} \\ 0 \\ \frac{1}{2} \\ 0 \end{bmatrix} \text{ and } \{q\}^{BC} = \begin{bmatrix} \frac{1}{2} \\ -\frac{1}{2} \\ \frac{1}{2} \\ -\frac{1}{2} \end{bmatrix}$$

Problem 8.11 (*9 points*) You are designing an aerial taxi, which uses four propellers to carry multiple passengers, and are interested in noise and power consumption. The vehicle's mass without any passengers is m_0, and each passenger is assumed to have mass m_p. When hovering, and carrying no passengers, the vehicle consumes P_0 power, and the propellers rotate at a speed Ω_0.

(a) Estimate the power required when the vehicle is carrying N passengers in hover.
(b) Estimate the propellers' rotational speed when carrying N passengers in hover.
(c) Updated safety regulations would require that each propellers' diameter be reduced by 10%. Compared to your answer for (a), how much more power is now required to hover with N passengers?

Problem 8.12 (*10 points*) In this problem you will compare the energy efficiency of a quadcopter, hexacopter, and octocopter. These vehicles consist of (respectively) 4, 6, and 8 propellers that are arranged symmetrically around the vehicle geometric center, as shown below.

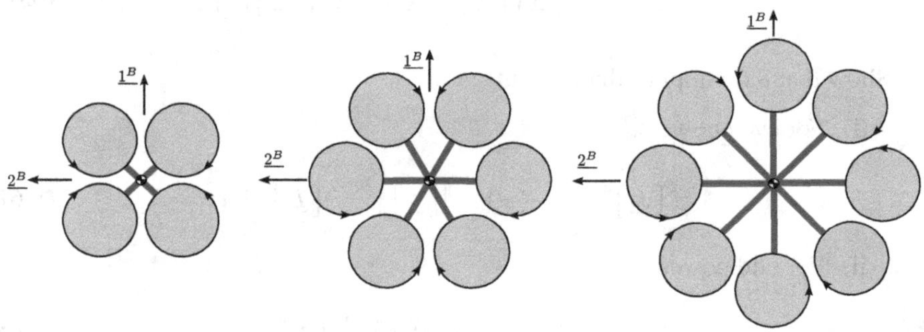

You are interested in a vehicle that can carry a payload of $m^P = 0.7$ kg, at the vehicle's geometric center. Each vehicle has a battery and electronics, located at the center of mass, with mass of $m^E = 0.3$ kg. Each motor has a mass of $m^M = 0.05$ kg, and is located at a distance r_m from the geometric center, with:

- quadcopter: $r_m = 0.15$ m
- hexacopter: $r_m = 0.22$ m
- octocopter: $r_m = 0.28$ m

The propellers have a radius $r_p = 0.1$ m, and can produce a maximum force of $f_{\text{max}} = 3.2$ N, and a minimum force of $f_{\text{min}} = 0.2$ N. You may assume that the propellers are massless, as is the frame and all other components.

Using the actuator disk model of propeller propulsion, predict the amount of power required at hover with each of the vehicles.

Problem 8.13 (*10 points*) For each of the vehicle types in Problem 8.12, compute the maximum possible angular acceleration about the body-fixed $\underline{1}^B$ axis, starting from rest. Approximate the vehicle's mass moment of inertia by treating all the propellers/motors as point masses, and the electronics/payload a single point mass.

Problem 8.14 (*10 points*) Consider a body B rotating with angular velocity $\underline{\omega^{BE}}$ relative to the inertial frame E, and at constant translational velocity $\underline{v_C^E} = \underline{0}$. What external moment $\underline{n_C}$ can be applied that maintain constant kinetic energy as defined in (8.50) (i.e., moments that do no work)? Find the most general expression.

You may use the following facts about first order tensors $\underline{x}$, $\underline{y}$:

$$\underline{K}(\underline{x})\,\underline{y} = -\underline{K}\left(\underline{y}\right)\underline{x}$$
$$\underline{K}(\underline{x})\,\underline{x} = \underline{0}.$$

References

Bar-Shalom Y, Li XR, Kirubarajan T (2001) Estimation with applications to tracking and navigation: theory algorithms and software. Wiley

Maruskin JM (2012) Introduction to dynamical systems and geometric mechanics. Solar Crest Publishing LLC

Mueller MW, D'Andrea R (2015) Relaxed hover solutions for multicopters: application to algorithmic redundancy and novel vehicles. Int J Robot Res 35(8):873–889

Mueller MW, Hehn M, D'Andrea R (2017) Covariance correction step for Kalman filtering with an attitude. J Guid Control Dyn 40(9):2301–2306

Shuster MD (1993) A survey of attitude representations. J Astronaut Sci 41(4):439–517

Simon D (2006) Optimal state estimation: Kalman, H infinity, and nonlinear approaches. Wiley

Stuelpnagel J (1964) On the parametrization of the three-dimensional rotation group. SIAM Rev 6(4):422–430

Zhang W, Mueller MW, D'Andrea R (2019) Design, modeling and control of a flying vehicle with a single moving part that can be positioned anywhere in space. Mechatronics 61:117–130

Zipfel PH (2007) Modeling and simulation of aerospace vehicle dynamics, 2nd edn. AIAA

Appendix
The Standard Atmosphere

The properties of the air vary with location, height, etc., but also with the season, and weather. For this reason, we refer to the 'standard atmosphere' as a representative of what typical atmospheric conditions will be. Below we tabulate some data, generated with the very easy-to-use Python `fluids` package (Bell and Contributors 2025). Note that this uses the '1976 US standard atmosphere', and the constant temperature (etc.) after 11 km is not a typo, but rather a feature of the model.

Height (z [m])	Temperature (T [K])	Pressure (P [kPa])	Density (ρ [kg m^{-3}])	Viscosity (μ [10^{-6}Pa s])	Speed of sound (v_{sound} [m s^{-1}])
0	288.2	101.3	1.225	17.894	340.3
1000	281.7	89.9	1.112	17.579	336.4
2000	275.2	79.5	1.007	17.260	332.5
3000	268.7	70.1	0.909	16.938	328.6
4000	262.2	61.7	0.819	16.612	324.6
5000	255.7	54.0	0.736	16.282	320.5
6000	249.2	47.2	0.66	15.949	316.5
7000	242.7	41.1	0.59	15.612	312.3
8000	236.2	35.7	0.526	15.271	308.1
9000	229.7	30.8	0.467	14.926	303.8
10000	223.3	26.5	0.414	14.577	299.5
11000	216.8	22.7	0.365	14.223	295.2
12000	216.6	19.4	0.312	14.216	295.1
13000	216.6	16.6	0.267	14.216	295.1
14000	216.6	14.2	0.228	14.216	295.1
15000	216.6	12.1	0.195	14.216	295.1
16000	216.6	10.4	0.166	14.216	295.1
17000	216.6	8.8	0.142	14.216	295.1
18000	216.6	7.6	0.122	14.216	295.1
19000	216.6	6.5	0.104	14.216	295.1
20000	216.6	5.5	0.089	14.216	295.1

For intuition: the City of Berkeley is at sea level ($z = 0$), and a typical passenger jet aircraft may cruise at around $z = 9000$ m (around 30,000 ft). Airports at high elevation include Denver Airport or Johannesburg's OR Tambo, at $z \approx 1650$ m (see,

M. W. Mueller, *Dynamics and Control of Autonomous Flight*, Mathematical Engineering,
https://doi.org/10.1007/978-3-031-91871-1

e.g., https://skyvector.com/). The international airport of La Paz, Bolivia, is at $z \approx 4000\,\mathrm{m}$.

A hot day in Berkeley may reach 40°C, so that the sea-level pressure is modelled (again, by the `fluids` atmospheric model) as $1.127\,\mathrm{kg\,m^{-3}}$.

Reference

Bell C, Contributors (2025) Fluids: fluid dynamics component of chemical engineering design library (ChEDL). https://github.com/CalebBell/fluids

Index

A
Acceleration, 51
Accelerometer, 73, 83, 89
Actuator disk model, 12, 143
Aileron, 14
Airfoil, 5
Angle of attack (α), 5
Angular momentum, 62, 151
Angular velocity, 22, 52, 159
Aspect ratio $\mathcal{R}$, 7
Atmosphere, 161
Attitude dynamics, 61
Axis-angle representation, 136, 153

B
Barometer, 79
Battery, 11
Bernoulli's law, 143

C
Camera, 82, 88
Cascaded control, 112, 114, 127
Center of mass, 59
Computer vision, *see* camera
Control, 105
Control surface, 14
Coordinates, 23
Coordinate system, 22, 31
Coordinate transformation matrix, *see* transformation matrix
Cross product, *see* vector product

D
Damping ratio, 110, 127
Density (ρ), 2, 161
Direction cosine matrix, *see* transformation matrix
Displacement, 22
Dot product, *see* scalar product
Drag, 1, 5
Dyadic product, 30
Dynamics, 59
Dynamic system, 105

E
Electronic speed controller (ESC), 11
Elevator, 14
Equilibrium, 106, 127, 148
Estimation, *see* state estimation
Euler angles, 34, 54, 142, 153
Euler's law, 63, 148
Euler symmetric parameters (ESP), 35, 139, 142, 153
Euler transformation, 55

F
Feedforward, 111, 124
Figure of merit, 13
First-order tensor, 27
Frame, 25, 31, 60
Free-stream velocity, 3

M. W. Mueller, *Dynamics and Control of Autonomous Flight*, Mathematical Engineering,
https://doi.org/10.1007/978-3-031-91871-1

G

Geometric control, 121
Global navigation satellite systems (GNSS), 77, 88
GPS, *see* Global navigation satellite systems (GNSS)
Gyro, *see* rate gyroscope

I

Identity tensor, 22, 27
IMU, *see* inertial sensor
Incompressible flow, 4
Induced drag, 8
Inertia, 148
Inertial frame, 60
Inertial sensor, 73
Inertia tensor, 22, 61, 69
Inner product, *see* scalar product
Input, 105
Integrator
  double, 109
  single, 108
Invariant description, 21
Inviscid flow, 4, 6

K

Kalman filter, 146
Kinematics, 49
Kinetic energy, 151, 159

L

Lagrangian mechanics, 64
Learning control, 121
Lidar, 82
Lift, 1, 5
Linear quadratic regulator, *see* LQR
Linear system, 107
LQR, 120
Lyapunov's direct method, 130
Lyapunov's indirect method, 118
Lyapunov stability, 116

M

Mach number (M), 4, 15
Magnetometer, 79
Mass moment of inertia, *see* inertia tensor
Matrix exponential, 137, 156
Mixer matrix, 97
Model predictive control, *see* MPC
Moment of force, 62, 159
MPC, 120

N

Natural frequency, 110, 127
Newton's law, 60
Noise, 146
Nonlinear system, 107
Norm, 23, 39, 41

O

Optical flow, 81

P

Parasite drag, 8
Pitch angle, *see* Euler angles
Position, *see* displacement
Principal axes of inertia, 62, 149
Propeller, 9, 93, 143

Q

Quadcopter, 93, 121
Quadrocopter, *see* quadcopter
Quadrotor, *see* quadcopter
Quaternions, *see* Euler symmetric parameters

R

Radar, 82
Range sensor, 80, 86
Rate gyroscope, 75, 83, 89
Rates, *see* angular velocity
Reynolds number (Re), 3, 15
Robust control, 120, 121
Roll angle, *see* Euler angles
Rotational time derivative, 49
Rotation tensor, 35
Rotation vector, 136, 156
Rudder, 14

S

Satellite, 152
Scalar product, 23, 29, 39, 43
Second-order tensor, 27
Sensor noise, 71
Sensors, 71
Skew-symmetric operator, 28, 42, 43, 57, 156
Speed of sound, 161

Stability, 115, 148
Stall speed, 15
State, 105
State estimation, 82

T
Tensor, 27
Thrust, 1, 9, 143
Tilt, 83
Torque, *see* moment of force
Trajectory generation, 124
Transformation matrix, 25, 32, 36, 135, 142, 153, 156
Translational momentum, 60

V
Vector product, 30, 40, 42
Velocity, 22, 51, 56
Viscosity (μ), 2, 161
Vision, *see* camera
Visual-inertial odometry (VIO), 88

W
Wing, 7
Wingspan b, 7

Y
Yaw angle, *see* Euler angles